Orchids

of Britain and Ireland

For Anne and Eleanor,
with thanks for their patience

A Pocket Guide to the *Orchids* of Britain and Ireland

Simon Harrap

B L O O M S B U R Y
LONDON • NEW DELHI • NEW YORK • SYDNEY

Acknowledgements

Much of the material in the present work derives from *Orchids of Britain and Ireland: A Field and Site Guide* (2nd edition, 2005) and I would like to reiterate the acknowledgements therein. In addition, thanks are due again to Sean Cole and Richard Gulliver for the use of their photographs of Irish Lady's-tresses, Robin Chittenden (Small-flowered Tongue Orchid), Bob Gibbons (Summer Lady's-tresses) and Nigel Redman (a British Ghost Orchid), as well as Mike Chalk for his lovely photo of Sawfly Orchid and Mike Waller for his similarly impressive shot of Coralrooot Orchids. The Botanical Society of Britain and Ireland again allowed use of maps from the *New Atlas of the British Flora*, Dr Andy Scobie shared his knowledge of Irish Lady's-tresses and Sean Cole and Mike Waller offered useful information and advice. Finally, Nigel Redman got the project off the ground and read through the entire text, making many useful suggestions, and Julie Dando advised on design.

Bloomsbury Natural History
An imprint of Bloomsbury Publishing Plc

50 Bedford Square
London
WC1B 3DP
UK

1385 Broadway
New York
NY 10018
USA

www.bloomsbury.com

First published 2016

British Library Cataloguing-in-Publication Data
A catalogue record for this book is available from the British Library.

ISBN: PB: 978-1-4729-2485-8
ePub: 978-1-4729-2486-5
ePDF: 978-1-4729-2487-2

2 4 6 8 10 9 7 5 3 1

Design by Simon Harrap
Printed in China by RR Donnelley

CONTENTS

INTRODUCTION

WHY THIS BOOK?

This pocket guide is designed to do two things: introduce wild orchids to a wider audience and show those who think they know orchids that there is always more to learn.

Few people realise that orchids are very much part of the natural heritage of Britain and Ireland. They range from the tiny, green Bog Orchid to the flamboyant Marsh Helleborine and the gorgeous Green-winged Orchid. They also include Lady's-slipper and Ghost Orchid, probably the two rarest native wild flowers in the British Isles.

All is not well, however, and orchids need friends. Despite the many schemes designed to mitigate its effects, modern industrialised agriculture, combined with urbanisation, insensitive forestry and an excess of plant nutrients (nitrates and phosphates, which encourage the growth of rank, coarse vegetation), has resulted in the ever-increasing homogenisation of the countryside. The beautiful, delicately woven tapestry of fields, pastures, woods and marshes created unwittingly by the hand of man over many generations is being put into the equivalent of a food blender that is reducing everything to a monotonous and anonymous wildlife desert. In the face of this assault, the majority of wild plants are in retreat, including all of our orchids. Do not be fooled by cheeky television presenters and glossy wildlife magazines, orchids and lots of other wildlife besides have disappeared from much of the landscape. If this book can help more people to enjoy, appreciate and value orchids and the places where they grow, it will have served its purpose.

WHAT IS AN ORCHID?

Orchids – the plant family Orchidaceae – are among the most diverse groups of plants, with over 1,000 genera and at least 25,000 species. Indeed, the Orchidaceae is probably the largest family of flowering plants (it vies for the title with the daisy family, the Asteraceae). The family derives its name from the Greek *orchis*, meaning 'testicle', a reference to the appearance of the underground tubers of some species. The term *orchis* was first used by Theophrastus (*c.* 370–285 BC) in his *Natural History of Plants*; he was a student of Aristotle and is considered to be the 'father' of botany.

Most people would recognise an orchid, even those without any particular interest in botany or gardening – at least the gaudy, hot-house hybrids and some of the more colourful wild orchids. Giving a precise definition is difficult, however, especially in non-technical terms, but European orchids share the following:

- They are perennial herbs rather than trees or shrubs, lacking woody parts.
- The leaves are simple and not divided into lobes or smaller leaflets.
- The leaves have no stalk and are arranged alternately along the stem.
- The flowers are carried in a single spike at the tip of the stem.
- The ovary is inferior, that is, placed *below* the sepals and petals.
- The male and female parts of the flower, the stamens and the stigma, are not separate but are fused together into a single structure called the column which lies in the centre of the flower.
- The flower is made up of three sepals and three petals but one of the petals differs from the others, usually significantly so, and forms the lip (sometimes known as the labellum). This is often brightly coloured and patterned, and intricately shaped. The lip is actually the uppermost petal but usually lies at the bottom of the flower because either the ovary or its stalk is twisted (the flower is therefore said to be resupinate).

ORCHIDS & FUNGI

The relationship that orchids have with fungi impacts on all aspects of their biology and, more than anything else, defines them.

Orchid seeds require fungi in order to germinate and grow. The seedling spends months or years underground and during this period is completely dependent on the nutrients that it obtains from fungi – it is 'mycotrophic' (a term deriving from the Greek *mukes* meaning 'fungus' and *trephein* 'to feed'). Seed must, however, be produced in large quantities to ensure that some, at least, will find the correct conditions for successful germination and growth, including the presence of the correct fungi. In turn, the need to produce large quantities of seed has powered the evolution of elaborate flowers and complex pollination mechanisms.

Even when the orchid has appeared above ground as an adult or near-adult plant and is able to photosynthesise and manufacture its own carbohydrates, in many species it still maintains a relationship with fungi. In a few orchids the adult plant continues to be entirely dependent on fungi (the so-called 'saprophytic' orchids). In others the adult plant is probably largely independent of fungi and gains its nutrients almost entirely from photosynthesis (these are 'phototrophic'). Most orchids, however, fall somewhere between these two extremes, with both sources of nutrition being utilised, perhaps in varying proportions depending on the season.

The ability to utilise two sources of nutrition allows orchids to thrive in marginal habitats; some grow in heavy shade and many are found on poor soils. In the tropics, orchids have extensively colonised the soil-less trunks and branches of trees and are epiphytes. Finally, the ability to fall back on fungi as a source of nutrition explains why many orchids are able to become dormant underground for a year, sometimes longer.

MYCORRHIZAS

It is thought that *c.* 90% of the world's plants have a relationship with fungi. Such a relationship is known as a mycorrhiza and the fungi that form these attachments are mycorrhizal. Mycorrhizal fungi live in the soil and, unable to manufacture their own carbohydrates by photosynthesis or obtain sufficient for their needs by the decomposition of organic matter, invade the root systems of green plants. But, rather than being parasitic, the fungus actually benefits the host plant by functioning as an extended root system, providing the plant with minerals, especially phosphorus; it may also confer some degree of drought, pest or disease resistance. The plant in turn provides the fungus with carbohydrates produced by photosynthesis; most plants are able to divert up to 20% of the carbohydrates that they manufacture via photosynthesis to their fungal partner without coming to harm. The relationship between the plant and the fungus is mutualistic (such mutually beneficial relationships were once termed 'symbiotic', but this term is now used for a wider range of interactions). Fungi form several kinds of mycorrhiza:

- Ectomycorrhizal fungi form a sheath or mantle over the plant's roots (the Greek *ectos* means 'outside'). Ectomycorrhiza are the dominant type of mycorrhiza formed by forest trees and are critical to their growth. Most ectomycorrhizal fungi are macrofungi and many produce recognisable mushrooms.
- Endomycorrhizal fungi (also known as VAM fungi) penetrate the cells of the plant's roots (the Greek *endon* = 'within'). They are microfungi and do not produce distinctive fruiting bodies. Poorly-known, they cannot be seen without a microscope and have proved impossible to cultivate and study in the lab.
- Ericaceous mycorrhiza are formed with the roots of various ericaceous plants.
- Orchidaceous mycorrhiza are formed with orchids.

ORCHID FUNGI

Until very recently, orchid mycorrhizas were thought to differ fundamentally from other mycorrhizal systems in that the orchid does not provide the fungus with carbohydrates: this must be true of orchid seedlings which develop entirely underground and have no green leaves. Rather, it is the fungus that provides the orchid with energy; there is good evidence that carbohydrates obtained by fungi from the decomposition of organic matter (or from trees, see below) is transferred to the orchid. In short, it seemed that orchids are parasitic on fungi (or, to use the terminology of scientific papers, they 'cheat' in their relationship with the fungus). There was no evidence for the transfer of nutrients from the orchid to the fungus, even in mature plants which are able to photosynthesise. Recently, however, it has been shown that carbon passes from Creeping Lady's-tresses to their associated fungi, and this has re-opened the debate regarding the orchid-fungus relationship.

The physical relationship between orchids and fungi is very sophisticated. Fungal hyphae pass through the outer layers of the orchid's root, rhizome or other underground organs and penetrate the cell walls to form loops and coils, called pelotons. At intervals, the orchid digests these pelotons and receives water, mineral salts, carbohydrate and other organic compounds from the fungus. So sophisticated is the orchid's use of fungi that it is able to control its spread and confine it to specialist cells; indeed, some orchids produce phytoalexins which act as a fungicide and prevent the fungi from reaching tubers and other storage organs. The main fungal associates of orchids are Basidiomycetes of the Rhizoctonia group (other members of the Rhizoctonia group are soil saprotrophs or pathogens).

'SAPROPHYTIC' ORCHIDS

Some orchids take the relationship with fungi to an extreme. Birdsnest, Coralroot and Ghost Orchids have no green leaves (or have the green pigments very much reduced) and throughout their lives depend entirely on their fungal partner for nutrition (they are fully mycotrophic). Furthermore, these species do not form associations with the usual orchid fungi of the Rhizoctonia group. It has been shown that both Birdsnest and Coralroot Orchids form relationships instead with ectomycorrhizal fungi which are simultaneously in partnership with nearby trees. Via these fungi the orchids acquire carbohydrates from the trees and therefore they are, in effect, parasitic on the trees. It has been suggested that forming associations with such ectomycorrhizal fungi may provide the orchid with a stabler and more reliable source of nutrients, which is particularly important when it has no other source of nutrition.

◀ Creeping Lady's-tresses

Birdsnest, Coralroot and Ghost Orchids are frequently but incorrectly described as 'saprophytic'. Saprophytes derive their nutrition from dead organic matter; these orchids acquire nutrients from living fungi.

Woodland orchids

It has also been shown recently that some other orchids, such as Red, White and Broad-leaved Helleborines, also form relationships with ectomycorrhizal fungi and are therefore able to utilise nutrients provided unwittingly by nearby trees. This may explain their ability to thrive in low light levels and to become dormant underground for long periods. Many are particularly associated with Beech trees.

Fungi and rare orchids

There is mounting evidence that some orchids are very specific about their fungal partners. Various orchids in the genera *Liparis*, *Goodyera* and *Spiranthes* have been shown to associate with just one species of fungus and to use the same species, both as germinating seeds and as adult plants. Other orchids form a relationship with many different fungi as adult plants but are very specific as germinating seeds (and may associate as seeds with a different species of fungus to any of those used by adult plants). Yet others seem to use a broad range of fungi for both germination and as adults, but a much more restricted group of fungi as underground seedlings.

Orchids that require a specific fungus in order to germinate and grow may be very limited in where they can live, compared to the more generalist orchids, and this has clear conservation implications. It seems likely that some of Britain and Ireland's orchids are rare and localised because their fungal partners are rare and localised. This may well explain why, for example, Burnt Orchid seldom colonises new sites. Any conservation programme, especially if it involves a reintroduction, may ultimately be unsuccessful if it fails to take this into account.

ORCHID SEEDS

Orchid seeds are rather small and, indeed, are often known as 'dust seeds'. They typically weigh 2–8 micrograms and, in British species, are 0.35–1.4mm in length. They are made up of a relatively simple embryo enclosed in a hardened carapace and surrounded by the much larger testa – a honeycomb of dead cell walls that traps air.

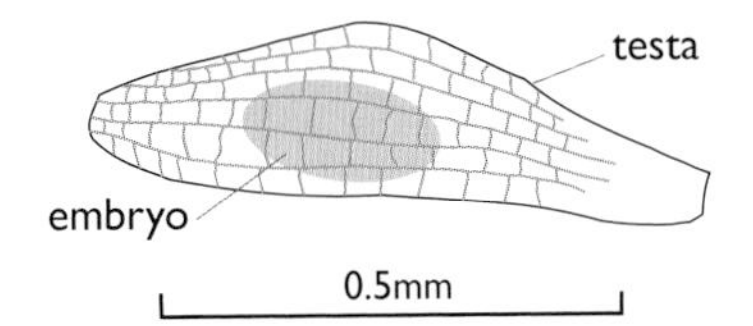

Small size confers several advantages. Large numbers of seeds can be produced at relatively little cost; in British orchids counts of between 376 and 25,000 seeds per capsule have been recorded (Lesser Twayblade and Greater Butterfly Orchid being the extremes). With many air spaces the seeds are ideally suited to wind dispersal and can travel long distances. They can also float on water, another effective means of dispersal. However, small size also imposes limitations. The seeds are so tiny that they contain very little in the way of food reserves (merely a few lipids and proteins) and they depend entirely upon fungi to provide their nutrients when they germinate.

Germination & Underground Growth

The orchid seed germinates 2–10cm below the surface of the soil. The seedlings are very vulnerable to desiccation and require the presence of fungi, so it is important for the seed to avoid germinating on the soil's surface. To ensure this, almost all orchid seeds will only germinate in darkness, once rain has washed them down into the soil or they have been covered by fallen leaves. Unlike other plants, they do not require light following germination because they do not produce green leaves.

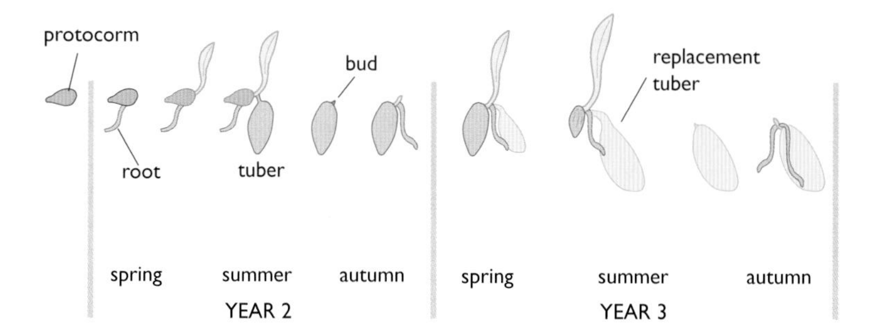

The pattern of growth from protocorm seedling to adult plant in the genus *Orchis*.

It is thought that most European orchids germinate in the spring (although there is only limited direct evidence for this) and therefore the seeds must have a mechanism to keep them dormant over the winter. Germination depends upon the uptake of water by the seed and the hard carapace forms a barrier against water, slowing or preventing germination for a while. In some orchids the carapace is incomplete and these seeds germinate much more easily and rapidly.

With its very limited reserves of nutrients, the orchid seed is dependent on fungi from the outset. In some species, such as Birdsnest Orchid, seeds start to germinate before infection by a fungus, although the breaking of the seed's dormancy almost certainly requires the fungus to be present in the immediate vicinity, and the seed probably responds to a chemical signal from the fungus. In other species, such as Coralroot Orchid, the seeds will only start to germinate after it has been infected by the specific fungal partner. From the outset the seedling is able to control the extent of infection by the fungus, confining it to certain areas (e.g. its roots) and prevent the fungus from turning the tables and invading its food stores.

Some species, including Birdsnest and Coralroot Orchids, form relationships with a very restricted range of fungi, perhaps just one species. Their seeds will only germinate in the presence of the appropriate fungus. (The implication is that the seeds can remain dormant for relatively long periods, waiting for the correct fungus to appear.) Other orchids form a relationship with a variety of common soil fungi that probably occur in all suitable habitats. Examples include the marsh orchids; their seeds will germinate and start to grow in the absence of the appropriate fungi, presumably because there is a very high probability that a suitable fungus will be encountered soon.

Upon germination the seed forms a protocorm. This is a small, often parsnip- or top-shaped structure with a scatter of root-hairs on its surface; these are single-celled projections that facilitate fungal infection. The protocorm usually goes on to develop roots and at this stage is often known as a mycorhizome. The primary function of these first roots is to host fungal activity rather than the supply of nutrients and water. In most orchids, as the seedling continues to develop, fungal activity is increasingly confined to the roots, and the mycorhizome, now free of fungal infection, is known as a rhizome. In some orchids the adult plant grows from a rhizome, but in many species the rhizome is largely replaced by tubers. In adult orchids, whether they grow from a rhizome or from a tuber, fungal activity is usually confined to the roots and, in some species, to the slender extremities of the tubers.

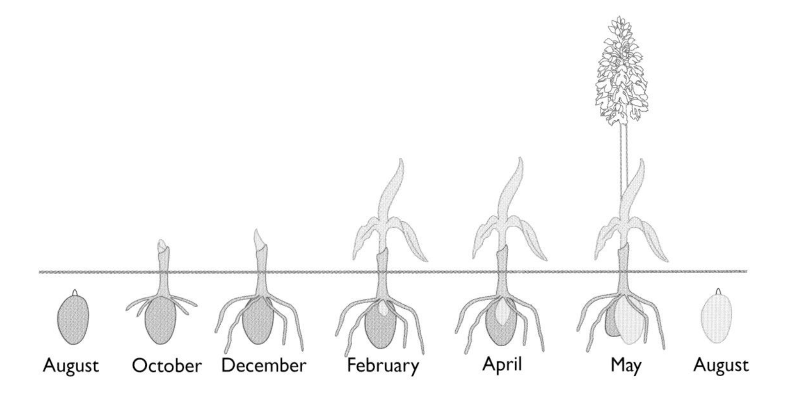

Annual growth pattern in the genus *Orchis*.

ANNUAL GROWTH PATTERN

The various genera of orchids display different patterns of growth, but the 'classic' pattern is shown by the genus *Orchis* (Early Purple Orchid etc.). At flowering time all *Orchis* species have two almost spherical tubers side by side at the base of the aerial stem. The 'tubers' are more accurately termed 'root-tubers' or 'root-stem tubers' and are essentially roots that have been modified to become specialist storage organs.

To follow the pattern of growth through the year it is best to start in the late summer when the orchid is 'resting' and consists only of a single tuber with one terminal bud; the leaves, rhizome and roots die off once the orchid has flowered.

In the autumn the bud on the tuber produces a short rhizome and from this a few roots develop. The roots serve two functions. First, their 'infection' with fungi provides the plant with nutrients. Second, they supply the plant with water and this becomes particularly important once the leaves have appeared. In some members of the genus the leaves develop in the autumn, in others they do not expand until the spring.

Once the roots have developed, they begin to produce nutrients. Some of these are stored in a new tuber that has started to form on the rhizome, side by side with the old; due to its relationship with fungi, the orchid can produce nutrients even before its leaves appear. The rhizome goes on to form the aerial stem and when the flowers open in the spring this has two tubers at its base (these two paired tubers, recalling testicles, gave rise to the name *orchis*). The older of these has supplied the current season's growth, including the leaves, stem and flower spike. This tuber is starting to shrink and will have vanished by the late summer's 'resting period'. The newer tuber is plump and swollen and continues to grow until the leaves die down. It will go on to overwinter and form the flower spike in the following year.

POLLINATION

Orchids are renowned for the beauty and complexity of their flowers. These flowers have not, however, evolved to amaze and delight us, but to fulfil the primary function of the plant, which is to reproduce itself.

Compared to most other flowering plants, orchids produce large quantities of seeds. Therefore large quantities of pollen have to be moved between the flowers because each pollen grain can only produce one seed. In most orchids the pollen grains are amalgamated together in large numbers to form pollinia. Due to their size and weight, these pollinia must be carried to other flowers by an insect and must be very securely attached if the process is to be completed successfully. The insect must be of the right size and shape, both to pick up the pollinium in the first place and to be positioned correctly in the next flower so that the pollinium makes contact with the stigma and effects fertilisation. European orchids are not usually pollinated with a whole pollinium, rather just fragments, and thus a single pollinium can, provided it remains on the insect, pollinate several flowers. But, if too small a quantity of pollen is deposited the capsule will mature without all the ovules being fertilised, and there will be many non-viable seeds.

Cross-pollination versus self-pollination

Most flowering plants reproduce sexually and the production of the next generation depends upon the successful unification of male and female gametes to produce seed; the male gamete is pollen and the female gametes are contained within the ovary. In most plants, including orchids, each flower produces both male and female gametes. It would seem a simple process therefore for pollen to be transferred within the same flower to the ovary. Indeed, why bother with the resources and energy needed to produce a flower at all when a small drab structure would be sufficient to bring pollen and ovary together? The answer lies in the advantages that cross-pollination brings to the species.

If pollen and ovary come from different individuals they will each carry a different set of genetic material and the resulting offspring will not, therefore, be genetically identical to its parents. This continual mixing of genetic material creates a large gene pool which gives the species flexibility and adaptability. It also means that individual mutations are disseminated through the population rather than being passed, unchanged, to the offspring; cross-pollinated species therefore tend to be relatively uniform in appearance.

If a flower is pollinated with its own pollen (i.e. it is self-pollinated) its offspring will have limited genetic variation. In turn, their offspring will be similarly limited. In self-pollinated orchids the overall genetic variation is rather low and the gene pool is limited, giving the species very little flexibility. Paradoxically, however, self-pollinated orchids can show more variation between colonies and are more likely to produce distinct varieties as mutations are more likely to be passed down unchanged to the next generation. Bee Orchid is perhaps the best example of this.

Bee Orchid var. *trollii* ('Wasp Orchid') ➤

There are circumstances where self-pollination is an advantage. If, for example, a single wind-blown seed produced a flowering Lizard Orchid many miles from other plants of the same species, it would have no chance of reproducing if it could only be cross-pollinated. If it is able to self-pollinate, it can produce seed and reproduce itself. Not surprisingly, self-pollination tends to be commoner in species that have scattered populations, or are at the edge of their range. Another factor may be a lack of pollinators, and species adapted to deep shade may well benefit from self-pollination; White Helleborine, an orchid that is routinely self-pollinated, is successful in Britain, whereas Red and Sword-leaved Helleborines, which are mostly or always cross-pollinated, have declined as increasing shade has made their woodland homes too dark for their pollinators. Self-pollination may therefore be a useful strategy in the short-term and it may only be over an evolutionary timescale, involving thousands of generations, that the advantages of cross-pollination come to the fore.

Most orchids are cross-pollinated but, rather cunningly, a large proportion have adaptations that prevent self-pollination but then allow the flowers to self-pollinate if, after a few days, a suitable pollinator has not come along. Conversely, the small number of routinely self-pollinated orchids are, at least occasionally, cross-pollinated by insects. In the world of orchids, nothing is straightforward.

INSECTS AS POLLINATORS

Orchid flowers have evolved to use insects to carry pollen from one plant to another and employ a variety of mechanisms to attract suitable pollinators. Bright colours and scents advertise their presence and the lip of the flower acts as a convenient landing platform. Many offer their insect visitors a reward of nectar but a large proportion do not. Their bright colours and scents are instead a *deceit* and they rely on the stupidity of insects, which are slow to learn that the flowers offer no reward. This deceit comes with a price, however, for Lady's-slipper and other orchids that offer no reward have consistently much lower rates of pollination than in orchids that do provide nectar.

Fly, Early Spider and Late Spider Orchids (genus *Ophrys*) have evolved an even more elaborate and deceitful mechanism – pseudocopulation – exploiting the sex-drive of insects with a combination of visual, olfactory and tactile deceits. In *Ophrys* the flowers have no nectar and offer their visitors no reward. The pollinators are usually bees and they are initially attracted to the flowers by chemical signals released by the lip of the orchid that mimic the pheromones produced by virgin female bees; the orchid produces such large quantities that the male bee may even prefer the flower's deceit to the real thing. The male bees home in on the orchid's pheromones, and as they get closer they catch sight of the flower and land on the lip. All three deceits now come into play. The male bee is still stimulated by the scent of the flower and the patterning on the lip but now also by the texture of the various hairy, velvety and smooth portions of the lip. This stimulates the bee to orientate himself into the 'correct' position, extend his genital apparatus and attempt copulation. Of course, the 'correct' position is the one in which he will pick up pollinia in the first flower visited and deposit the pollinia on the stigma of the second flower and so on. Eventually this process of pseudocopulation is interrupted and the bee leaves.

The orchid is totally dependent on the bee. If there is no pollination, there will be no seed and therefore no more orchids; a powerful engine to drive the evolution of the orchid. For the orchid, the better and more precisely it mimics its pollinator the better its chance of successful pollination. Each species of orchid therefore evolves to mimic the pheromones of a particular species of bee in order to maximise its chances of attracting pollinators. In this way the genus *Ophrys* has evolved into many species, reflecting the many species of potential pollinator. Each one is precisely matched.

Paradoxically, and by contrast with all other members of the genus *Ophrys*, Bee Orchid is routinely self-pollinated. It had clearly evolved to use the same mechanism as the other species, with an elaborate pattern and texture to its lip, but has largely abandoned cross-pollination, presumably relatively recently. The short-term gains of self-pollination are evident, as it is by far the commonest and most widespread of the four British *Ophrys*.

ORCHID CONSERVATION

Orchids face three major threats: habitat destruction, habitat change and human predation.

Habitat destruction has clearly taken the greatest toll. Farming, forestry and other developments have destroyed innumerable orchid sites, especially in the period since World War Two. Most of the destruction has been state-sponsored through the operations of the Common Agricultural Policy and the Forestry Commission. Between 1945 and 1980 the Forestry Commission attempted to destroy and re-plant with conifers 200,000 hectares of ancient woodland, to say nothing of the tens of thousands of hectares of heathland, moorland and sand dunes that were destroyed. The CAP has been reformed in recent years and attitudes and policies at the Forestry Commission have changed (although there is still a great reluctance to undo much of the damage done in the name of near-worthless timber). But, despite much lip-service in recent years, few politicians have any commitment whatsoever to conservation and when push comes to shove development almost always takes precedence over wildlife. No wonder the government conservation agencies (Natural England, Natural Resources Wales and Scottish Natural Heritage) have often been accused of being lame ducks in the face of their political masters, despite the best efforts of their staff. In terms of habitat destruction, the prospects for orchids remain bleak.

Habitat change has only recently been acknowledged as a major issue. It has come to be recognised that orchids do not live in stable, 'climax' communities of plants, at least in the British Isles, rather in habitats that were created and maintained,

Burnt Orchid has gone from *c.* 80% of its historic range, due both to habitat destruction and habitat change.

albeit inadvertently, by people. Grassland, marshes, heathland and woodland are all the product of traditional land-use. Once these traditions died out, habitats started to change, slowly at first but then rapidly, and many have become unsuitable for orchids. Ironically, this applied especially to reserves, where a fence and a 'keep out' sign were often the limit of any management. Now, conservationists try to replicate the traditional land-uses, often at great expense, that created and maintained the habitats they manage. In lowland Britain the reinstatement of grazing is often the single most important measure that can be taken to help orchids.

Human predation has often been seen as a major threat to orchids, be it innocent ramblers picking bunches of flowers or avaricious botanists determined to get another specimen for their collections or gardens. More recently, photographers and even visitors keen to merely look at plants have joined the list of 'threats'. The answer has traditionally been secrecy, and details of the locations of the greatest rarities were and still are jealously guarded; even the location of huge colonies of species such as Burnt and Early Spider Orchids was veiled in secrecy.

Human predation certainly poses a threat to those species that occur in such small numbers that a significant part of a population (or even the whole population) can be stolen. There are still cases where plants are dug up illegally, from Bog Orchid to Lizard Orchid. Perhaps the most notorious in recent years was the attack on the single Lady's-slipper growing in Silverdale in Lancashire. Natural England had made the bold decision to allow limited publicity and a large number of people had been able to admire this beautiful orchid. The fact that this particular plant was probably originally of garden origin does not in any way lessen the damage done.

Despite the odd incident, however, for most orchids human predation is, in the final analysis, irrelevant to their fortunes, especially in the face of habitat destruction and habitat change. Unnecessary secrecy has indeed probably led to the destruction or degradation of many sites, as those responsible for the land remain in ignorance of its importance. It has also deprived many people of the enjoyment of seeing the orchids and many potential friends for orchid conservation have surely been lost in this way.

There have been a few special conservation initiatives involving orchids. In 1983 the Sainsbury Orchid Conservation Project was established at the Royal Botanic Gardens, Kew. This has involved research into the propagation of orchids with a view to reintroducing some of the rarer species. A range of orchids has been involved, including Military and Fen Orchids, with a substantial effort going into the reintroduction of Lady's-slipper. Reintroductions are controversial, however, with some conservationists arguing that the time and effort could be better spent on conserving existing populations. A second initiative at the Royal Botanic Gardens was the establishment in 1997 of the Millennium Seed Bank. This is intended to store viable seeds for as many of the world's plants as possible, including, of course, British wild orchids. Techniques have been developed which should allow orchid seeds to be stored for long periods, although some species cannot yet be cultivated successfully.

WHAT YOU CAN DO

The first step in orchid conservation is accurate and up-to-date information on their distribution and abundance. Amateur botanists provide the vast majority of information on plant distribution in Britain and Ireland via the system of county recorders organised by the Botanical Society of Britain and Ireland (BSBI); they are always pleased to receive records, with details of the species involved, numbers, date and location. There is also always a need for volunteers to undertake practical habitat management on reserves and other sites; the local wildlife trusts are the first contact if you are keen to get involved.

The greatest contribution individuals can make to orchid conservation is, in our opinion, to become a 'local champion'. Getting to know an area intimately, finding and recording orchids and other wildlife, and then badgering local councils, wildlife trusts, government agencies or church-wardens to sit up and do what is necessary to safeguard the good areas. This may not make you popular in some quarters but may, in the end, get things done.

ORCHIDS & THE LAW

All orchids and, indeed, almost all wild plants, are protected in Britain by the Wildlife and Countryside Act, 1981 and cannot be uprooted unless you are the owner or occupier of the land or have their permission to do so. In addition, some of the rarer orchids enjoy much greater protection under Schedule 8 of the Act and it is illegal for anyone, even the owner or occupier of the land, to uproot, destroy or pick these orchids. The term 'pick' is defined to include gathering or plucking any part of the plant, including collecting seeds. It is also illegal to posses any live or dead wild plant in Schedule 8, or any part of or anything derived from such a plant, or to trade in such items.

ORGANISATIONS TO JOIN

Botanical Society of Britain and Ireland (BSBI)

www.bsbi.org.uk. The learned society for professional and amateur botanists. Publishes the *New Journal of Botany* as well as the less formal (and very readable) *BSBI News*, organises a network of county plant recorders, runs field meetings and has a panel of referees to advise on identification. Notably, the BSBI website is a gold mine of useful information and resources. In short, the BSBI is the society for anyone with a keen interest in wild plants.

Plantlife International

www.plantlife.org.uk. The wild plant conservation charity. Has a small number of reserves, runs 'back from the brink' projects for many declining species and promotes various surveys to raise awareness of wild plants and their conservation. Publishes a quarterly magazine, *Plantlife*.

Wild Flower Society

www.thewildflowersociety.com. Established in 1886 for amateur botanists and wild flower lovers in the UK. Organises meetings to see and photograph British wild plants in their natural habitats.

NOTES ON THE SPECIES ACCOUNTS

NAMES

The English and scientific names used in the text and the order of the species accounts largely follows the *New Flora of the British Isles* by Clive Stace (3rd edition, 2010). A few alternatives are given. Scientific names reflect the relationships between species and in recent years new evidence, especially from genetic studies, has greatly improved our understanding of these. This is reflected in new scientific names for some species.

FLOWERING PERIOD

A guide is given to the period in which the species is likely to be in flower. Flowering times do vary, however, both predictably and unpredictably:

* The weather over the preceding weeks and months will affect flowering times – most obviously in early-flowering orchids.
* Orchids growing further to the north and at higher altitudes tend to flower a little later (although there is often surprisingly little difference between S England and Scotland).
* Orchids growing in wetter habitats

- ✻ will flower later than the same species growing in drier habitats.
- ✻ Orchids growing on or very near to the coast will tend to come into flower a little earlier than those inland.
- ✻ Orchids growing on sheltered south-facing slopes will flower earlier than those with an exposed, westerly aspect or those facing north.

There can also be marked and unpredictable variations between colonies, even those close to each other and, in recent years, perhaps as a result of 'global warming', many orchids have been coming into flower earlier; it is worth bearing this in mind if you are hoping to see a species at its best.

HABITAT

The information given in this section applies to Britain and Ireland; in some cases orchids occupy a rather broader range of habitats elsewhere. Most maximum recorded heights above sea level come from the *New Atlas of the British & Irish Flora* (2002).

RANGE MAPS

Accompanying the text is a range map, where the distribution is given by 10km squares. Each dot represents the presence of the species in the period 1987-1999, corresponding to the period when records were collected for the *New Atlas of the British & Irish Flora*. Although now a little dated, these maps, which were meticulously checked, represent the best and most accurate picture of the current distribution of orchids in Britain and Ireland.

DEVELOPMENT & GROWTH

Orchids are notorious for the wide variations from year to year in the number of plants in flower. It used to be thought that this was related to fluctuations in the size of the population and that some species, such as Bee Orchid, were monocarpic and therefore flowered just once before dying. It is now known that most orchids are relatively long-lived and the total population, including non-flowering plants and those dormant underground, is often fairly stable.

Fluctuations in the number of plants flowering are related to growing conditions both in the current year and in the previous growing season (which may be either the previous summer or the previous winter, depending on the species). Growing conditions are, in turn, usually related to rainfall. Wet weather is conducive to growth but prolonged dry spells can be very bad for orchids and in some cases can severely restrict flowering.

For some species long-term research has resulted in information on the longevity of orchids. This is usually expressed as the 'half-life', which is a measure of the life expectancy of an orchid after its first appearance above ground. It marks the point at which 50% of the population that emerged in any given year has died.

Status & Conservation:

CONSERVATION DESIGNATIONS

The latest UK conservation designations are included. 'Nationally Scarce' denotes that the species has been recorded from 16–100 10km squares in Britain from 1970 onwards (the designation does not include Ireland). 'Nationally Rare' includes all plants recorded from 15 or fewer 10km squares in the period from 1987 onwards. Each species is also assigned to a threat category: Critically Endangered, Endangered and Vulnerable indicate decreasing levels of threat, while Data Deficient denotes a species where there is inadequate information on its distribution and/or population status to make an assessment of the risk of extinction, but where it is thought that it may be threatened.

Description

A detailed description of the orchid is given, taken in many cases from the living plant. I have depended on the literature for some details, especially measurements. It is worth remembering that many orchids can be taller than the range quoted.

SUBSPECIES

Some orchids show definite patterns of variation which may be either geographical or ecological; plants in a particular region or a particular habitat may differ consistently in appearance from other areas. In these cases I have recognised different subspecies. There is much disagreement about which subspecies are worthy of recognition and I have not followed any one authority in this.

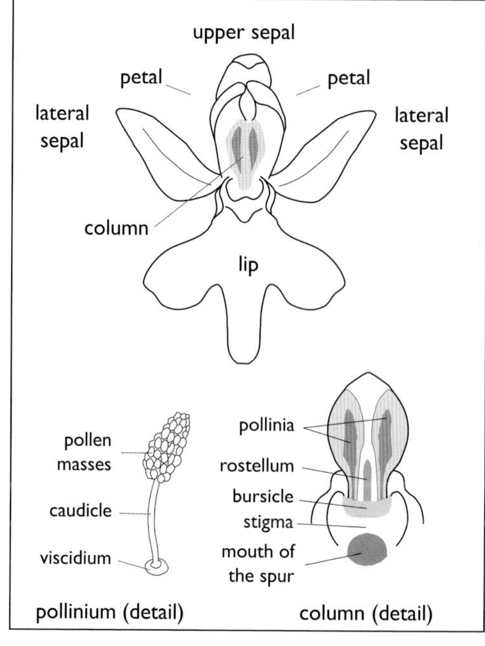

VARIATION

Brief details are given of the normal range of variation that can be encountered in the species, as well as details of some of the named varieties (usually abbreviated 'Var.'). Like all organisms, orchids are subject to random mutations of their DNA which may produce a variety of aberrant and deformed plants, and, like stamp collectors (and the butterfly collectors of old), orchidologists have traditionally been fascinated by these abnormalities. I do not share that obsession and consider that the odd plant that produces flowers that are deformed or unusually coloured in some way to be of little special interest. However, some deformities shed light on more primitive stages in the orchid's evolution. In addition, species such as Bee Orchid throw up the same mutation again and again in widely separated localities, and in such cases these 'varieties' (var. for short) are often given names. Unlike subspecies, varieties do not show definite geographical or ecological patterns and can pop up anywhere.

HYBRIDS

Hybrids have, like varieties, long fascinated orchidologists and in some cases are either very attractive or throw light on the relationships between orchids. Hybrids between species in the same genus tend to occur much more frequently than hybrids between species in different genera and, for example, the frequency with which Frog Orchid hybridised with the spotted and marsh orchids was a long-standing clue to its close relationship with those species. On the other hand, many hybrids are undistinguished and their true parentage, and even their status as a hybrid, may be the subject of guesswork, sometimes highly ambitious guesswork. All too often, plants are diagnosed as hybrids when they are merely aberrant individuals (or even within the range of normal variation). I have usually only included hybrids that are listed in the *Hybrid Flora of the British Isles* (2015) and have given the names used in that work.

LADY'S SLIPPER *Cypripedium calceolus*

IDENTIFICATION

One of Britain's rarest wild flowers, with just one cluster of plants of native origin surviving at a site in Yorks. Happily, a successful re-introduction programme means that Lady's Slippers can be seen in flower at several sites in Yorks and Lancs. Height usually *c.* 30cm and rarely more than 60cm. Unmistakable. The specific scientific name *calceolus* means 'little shoe' and, like the English name, refers to the slipper-like appearance of the lip. **SIMILAR SPECIES** None. **FLOWERING PERIOD** Late May–early or mid June. Plants are in flower for 2–3 weeks. Each flower lasts 11–17 days, withering on the sixth day after pollination.

HABITAT

Prefers relatively well-lit areas but likes to have its roots in cool, moist soil. The surviving native plants grow in species-rich grassland on a fairly steep, well-drained, north-facing slope in a sheltered limestone valley. Former English sites were in ash, hazel and oak woods on steep, rocky slopes, always on limestone – recalling the open woodland favoured in Europe.

POLLINATION & REPRODUCTION

Pollinated by small bees, especially in the genera *Andrena* and *Lasioglossum*. The flowers do not produce nectar and bees are probably attracted by the flower's scent, which may mimic the bees' pheromones (chemical signals that are associated with feeding and mating behaviours), while the colour and markings of the flowers are probably also important in the deception.

A bee lands on the edge of the slipper's upper opening or tries to land on the staminode (see below), and then falls into the slipper. After a few minutes it tries to leave. The sides of the slipper are very smooth and slippery and the rim curls over and inwards, making escape impossible. The bee can only leave through the small openings on either side of the column where there are small stiff hairs to give it a foothold. These openings are only just big enough for the bee, which is forced into contact with one of the stamens as it makes its escape, picking up a load of pollen. The bee goes on to visit another flower, and when in turn it eventually leaves this its back rubs against the stigma, which projects down into the slipper, and pollen from the first flower is deposited there – the surface of the stigma has minute, stiff, pointed papillae that act as a brush to remove pollen from the bee's back. As it escapes, more pollen is carried away, ready to be deposited on the next flower and continue the process.

The mechanism is precise and the bee has to be the right size; bees that are too large or too small can escape without pollinating the flower. A wide variety of other insects also enters the slipper but these too are the wrong size and shape and either leave unharmed or may be trapped and die. Self-pollination is unlikely; the bee would have to reverse back into the flower just as it was on the point of escape. In addition, it seems that the flowers are, to a great extent, self-sterile.

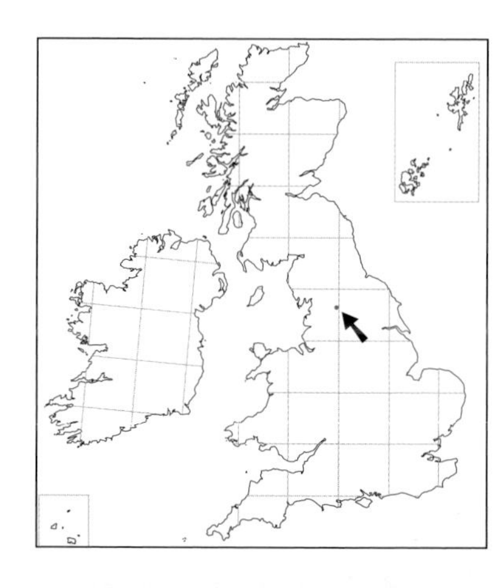

The pollination strategy is not efficient and seed set is rather poor with few fertile capsules produced. Bees are attracted to large groups of flowers, especially those in sunlight, but even in large populations in Europe an average of just 10% of flowers set seed. Nevertheless, each capsule contains 6,000-17,000 seeds which may be dispersed by rain – the seedpods seem to close up when dry and open when wet.

Reproduces vegetatively through division of the branching rhizome, and in many populations in Europe this is thought to be more important than seed in the recruitment of new plants.

DEVELOPMENT & GROWTH

The first green leaves are reported to appear in the fourth year after germination by some authors and in the first year by others. The immature plants have a slender stem with1–2 small leaves and may remain in this state for several years. In England a seedling has been noted to flower nine years after it first appeared above ground, and in Europe the young plant takes 6–10 years to produce flowers. Plants are long-lived; many are over 30 years old, with some over 100 years; a life span of 192 years has been determined from the examination of a single rhizome in Estonia.

STATUS & CONSERVATION

Nationally Rare and listed as Critically Endangered: WCA Schedule 8. Formerly found widely but locally in the limestone districts of N England, from Derbyshire to Co Durham, with most records from W and N Yorks. Lady's-slippers were, however, obvious subjects of curiosity and were picked or dug up from at least the 16th century onwards. In the late 18th and early 19th centuries they were ruthlessly stripped from the wild for horticulture and as herbarium specimens and by the mid 19th century the species was rare. In 1917 it was declared extinct in Britain.

In 1930, Lady's-slipper was resurrected from the dead when a single plant was found in a remote Yorkshire dale (this plant has survived to the present day). However, the last note of its rediscovery in print was in 1937, and the species slipped from the botanical world's attention. And, from 14 stems and one flower in 1930 it dwindled to 2–5 stems by the late 1940s and 1950s and seldom flowered: single blooms were produced in 1934 and 1943 but not again until 1959.

Despite the secrecy, in the 1960s word started to get out and the Lady's-slipper faced the old threat from collectors and a new threat from visiting botanists with big feet and heavy cameras. Indeed, the site was raided and half the plant was removed. The 'Cypripedium Committee' was formed in 1970, with representatives from various conservation and botanical interests. Its first priority was to safeguard the sole remaining wild plant, and this has been guarded every year since then – potential visitors are asked to keep away.

With careful protection and habitat management the Yorkshire Lady's-slipper has slowly increased in vigour, with a steady increase in the number of shoots and flowers on the main clump (which may be just one plant, several clones or even include seedlings). Few or no flowers were being pollinated naturally, however, and hand-pollination began in 1970. This resulted in good seed set and the production of many seed capsules; some are left to mature on the plant while others have been sent to Kew Gardens (two plants taken from wild sites in the early 20th century have survived and are used to cross-pollinate the Yorkshire plant).

As part of the 'Species Recovery Programme', organised by Natural England, research began in 1983 at Kew. The fungus that aids the Lady's-slipper seeds in germination and growth could not be identified, however, thwarting efforts to cultivate plants from seed. After much trial and error, a method of germinating seed in the absence of fungi was developed. This involves supplying the nutrients

directly to the seedling in a sterile medium. Although only *c.* 10% of seeds germinate, large numbers of seedlings can be produced. By 2003 *c.* 2,000 seedlings had been planted out in 23 locations (17 still had plants in 2010). Survival was not good, however, with slugs and snails a particular problem; 5-year-old-plants have been found to be the best at establishing themselves. The first re-introduced plant flowered in 2000, 11 years after being planted out. In 2009, seed pods formed after natural pollination by insects and by 2015 good numbers of re-introduced plants were flowering, with Gait Barrows NNR near Silverdale, Lancs, the best-known and most accessible public site.

A single plant has also been present at Silverdale for many years, although it is thought that it was planted there in the late 19th or early 20th century; its DNA suggests that it is from either Austria or possibly the Pyrenees. It did not flower for many years but slowly increased in vigour and by 2004 produced nine flowers. Sadly, later in the 2004 season this plant was vandalised and probably partially removed. It survived this thoughtless attack, producing seven flowers in 2009, but by 2015 was reported to be in decline.

DESCRIPTION

UNDERGROUND Grows from a slender, creeping, branched rhizome. Each branch may eventually put up an aerial stem; as a plant ages the number of flowering shoots increases. **STEM** Glandular-hairy with 3-4 green or brown sheaths at the base. **LEAVES** 3–4 (–5), arranged alternately up the stem; oval, elongated to a pointed tip, wavy edged and very prominently veined; sparsely hairy (especially on underside), ciliate along margins. **SPIKE** Each stem usually produces 1–2 flowers, very rarely three. **BRACT** Leaf-like, longer than flower and held erect behind it. **OVARY** Long, slender, 6-ribbed, curved (but not twisted), with glandular hairs and a short stalk. **FLOWER** Sepals purplish-brown or claret with wavy edges, downy on inner surface and hairy at base; upper sepal erect, lanceolate; two lateral sepals fused, hanging vertically below lip (their tips forming two small teeth at the tip of the combined 'synsepal'). Petals purplish-brown, mottled olive-yellow towards base, with downy midribs and long hairs at base; strap-shaped, pointed, variably twisted (through up to 360°) and hanging on either side of lip. **LIP** Yellow, resembling a bag, with a large entrance on upper side towards rear and two small openings on either side of column at base. Edges of upper opening rolled down and under, interior of slipper covered in sticky hairs with lines of reddish dots along its floor. **COLUMN** Projects forwards into slipper, divided into two parts: the staminode (yellowish-white variably marked with red spots and very conspicuous) and the large fleshy stigma on the lower part of the column hidden inside the slipper. Two other stamens lie on either side of the base of the staminode, adjacent to two small rear openings into the slipper. **SCENT** Delicate, said to be sweet, recalling oranges. **SUBSPECIES** None. **VARIATION** None. **HYBRIDS** None.

WHITE HELLEBORINE
Cephalanthera damasonium

IDENTIFICATION

Relatively common and widespread in S and SE England. Height 15–40cm (8–67.5cm); many are just 13–18cm with only 1–2 flowers. The loose spike of *upward-pointing creamy-white flowers* is very distinctive. The flowers do not normally open widely, the sepals, petals and lip forming an *egg-shape* around the column, but in a few plants the lateral sepals spread apart like outstretched arms and the upper sepal and petals then form a loose hood. Most flowers set seed, and the stout, elongated capsules, held *upright*, can identify the species well after the flowers have withered. **SIMILAR SPECIES** Sword-leaved Helleborine is much rarer but is sometimes found together with White Helleborine. The flowers of Sword-leaved Helleborine are always pure white and tend to open more widely. In addition, the sepals have more pointed tips, and the leaves, especially the lower ones, average rather longer and narrower. The most reliable distinction is the length of the bracts: on Sword-leaved Helleborine these are *shorter* than the ovary, at least in the upper part of the spike (on the lowest 2–3 flowers the bracts may be long and leaf-like); on White Helleborine the bracts are *longer than the ovary on all flowers*. **FLOWERING PERIOD** Mid May–late June, exceptionally late April–mid July; flowers earliest in the open or in very dry woods.

Very infrequently the flowers open widely, revealing the yellow pseudopollen, the raised ridges on the lip, and the long column projecting into the flower.

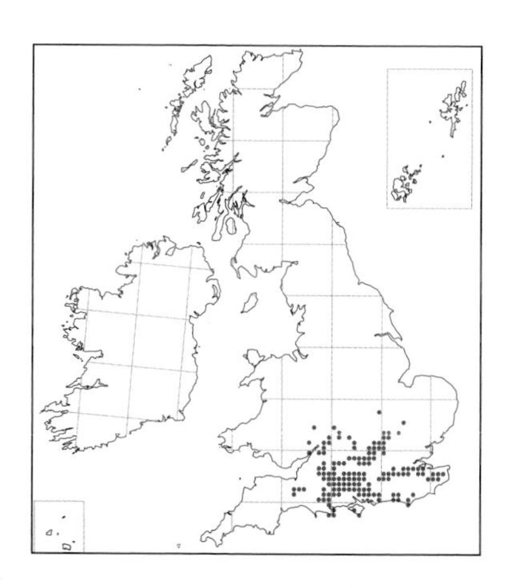

HABITAT

Deciduous woodland and shelter belts, sometimes scrub and occasionally nearby grassland (especially on north-facing slopes), always on well-drained soils over chalk or limestone. Strongly associated with Beech, even solitary trees, and will grow in the dense shade cast by that species (indeed, sometimes almost the only flowering plant, growing through a carpet of dead leaves or on bare stony or mossy ground). The most robust plants are, however, found where the shade is not too intense. May colonise newly available habitats, such as maturing beech plantations.

POLLINATION & REPRODUCTION

Probably largely self-pollinated. The anther opens while the flower is still in bud and releases the pollinia, which are very friable; at least some of the pollen adheres to the stigma and effects pollination. Once the flower is mature, however, the outer part of the lip folds down to form a landing platform for insects, especially bees. These are attracted by the golden-yellow mass of pseudopollen on the tip of the lip. Visiting insects may not only spread more pollen onto the stigma within the flower but also sometimes transfer pollen from flower to flower, thus causing cross-pollination. But, once the flower is fully fertilised, the lip folds up again to 'shut the door'. Pollination is very efficient and almost all flowers produce seed. Vegetative propagation also occurs – the roots produce buds that develop into new aerial shoots.

DEVELOPMENT & GROWTH

Mature plants acquire the majority of their nitrogen and roughly half their carbon from a fungal partner, which explains the ability of White Helleborine to thrive in densely shaded sites. Although it appears to form relationships with a wide range of fungi, White Helleborine preferentially forms associations with ectomycorrhizal fungi, especially Basidiomycetes; these in turn have a relationship with nearby trees. The orchid draws nutrients from the roots of these trees via the fungi but 'cheats' the fungus by giving nothing in return. Albino, chlorophyll-less White Helleborines (var. *chlorotica*) are sometimes found, and these must be fully mycotrophic and completely dependent on fungi.

STATUS & CONSERVATION

Listed as 'Vulnerable'. Relatively common within its limited range and can occur in large numbers. White Helleborine has, however, been lost from over 40% of its total historical range, largely due to woodland clearance and the replanting of woodlands with conifers. Many of the losses have been on the edges of the range.

DESCRIPTION

UNDERGROUND The aerial stem grows from a deeply buried, short, woody rhizome. **STEM** Single, green, the upper part ridged and either hairless or slightly hairy. 1–3 brownish, membranous sheaths at the base, the uppermost sometimes tipped green. **LEAVES** 3–5, placed alternately up stem, more or less in two opposite rows, curving gracefully upwards to lie horizontally; oval to broadly lanceolate, tapering to a moderate point and becoming narrower and more bract-like towards spike; lowest leaf rather short and cowl-like; greyish-green, sometimes with a bluish tinge, with prominent veins (especially on underside). **SPIKE** Loose, with 1–12 relatively large flowers, sometimes as many as 16, most pointing vertically upwards. **BRACT** Greenish, narrowly lanceolate and relatively long, often much longer than the flower, but becoming shorter towards tip of spike (although still longer than ovary). **OVARY** Pale green, either not twisted or only moderately so, slender, cylindrical and boldly 6-ribbed. **FLOWER** Sepals white to creamy-white with a hint of green, tear-shaped to oval-lanceolate, the broader end at the base, tip blunt. Petals slightly shorter and more oval. **LIP** Short and broad, divided into inner (hypochile) and outer (epichile) halves by a distinct narrowing or 'waist'. Hypochile and base of epichile held parallel to the column, with sides curved upwards and inwards to form a gutter, the base of which is washed golden-yellow. Epichile heart-shaped, rather broader than long, the tip curving gently downwards; towards the tip an extensive patch of golden-yellow frosting with 3–5 longitudinal ridges and furrows towards its base; these ridges are also washed golden-yellow. **COLUMN** Long, slender and whitish, anther tinged pale yellow. **SCENT** None. **SUBSPECIES** None. **VARIATION** **Var. *chlorotica*** is deficient in chlorophyll, with very pale green or even yellowish-white stem and leaves. Very rare. **HYBRIDS** ***C.* x *schulzei***, the hybrid with Sword-leaved Helleborine, is very rare. Recorded from Hampshire and West Sussex.

SWORD-LEAVED HELLEBORINE

Cephalanthera longifolia Other names: **Narrow-leaved Helleborine**

IDENTIFICATION

Widespread but rare and very local. Height 15–65cm (exceptionally 5cm in very exposed situations). Distinctive, with long, gently arching leaves alternating up the stem and spires of *pure white flowers*. The bracts are *shorter than the ovary, at least in the upper part of the flower spike* (they may be long and leaf-like on the lowest 2–3 flowers), and the flower spike is clearly demarcated from the leafy part of the stem. The vast majority of flowers do not open widely, the sepals and petals cupping the lip with the tips of the sepals flared outwards just enough to display a large golden-yellow patch on the lip. On some, however, the lateral sepals are held spreading. **SIMILAR SPECIES** See White Helleborine (p.25). **FLOWERING PERIOD** Early May–mid June. Can vary by 2–3 weeks from year to year, but often at its best in the last two weeks of May, even in Scotland; sometimes still in flower in early July at exposed sites in the N.

HABITAT

Optimum sites are at the interface of woodland and grassland in places that are sunny for at least part of the day but where the ground vegetation is not too dense – glades, clearings, rides and the margins of roads and tracks. It can persist in dense shade or under scrub, but flowering will be very much reduced, the plants remaining in a vegetative state or even becoming dormant underground. It can, however, flower spectacularly if the scrub is cleared and the light intensity rises above a certain level.

In S England, Sword-leaved Helleborine favours beechwoods on chalk. Intriguingly, in Hants, rather than being in ancient woodland, many sites are along ancient trackways or in secondary woods that were arable fields in the 19th century but then allowed to revert to woodland. Elsewhere in Britain it grows under a variety of deciduous trees – in the Wyre Forest in Worcs it once thrived in coppice cut on an 18-year cycle (the long coppice cycle allowing time for it to build up resources to flower) and it has also been found in pine plantations at Newborough Warren on Anglesey. Occasionally occurs on chalk grassland, but the plants are very small, with few flowers.

Many sites for Sword-leaved Helleborine are on calcareous soils overlying chalk

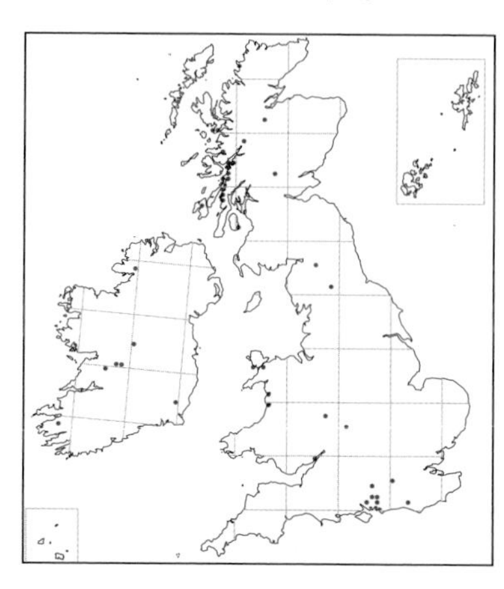

and limestone, but in Wales it is found in woodland on more acid soils. Tolerant of both wet and dry conditions, from damp woods or wet scrub in the N and W to dry chalky slopes in S England.

POLLINATION & REPRODUCTION

Pollinated by solitary bees, mostly in the family Halictidae, which are typically found in flower-rich grassland and visit the sunnier parts of adjacent woodland to forage. Sword-leaved Helleborine produces no nectar and relies on deceit to attract its pollinators; the golden-yellow pseudopollen on the lip may be especially important in luring insects. However, for the deceit to function effectively there must be a good supply of genuinely nectar-producing flowers in the vicinity to attract the bees to the area. Furthermore, bees will only visit flowers in bright sunlight and forage mostly 10am–3pm, and both the helleborines and the surrounding nectar-producing plants have to be in sunlight for some of that period; sites that catch the morning sun may be especially favoured. Suitable sunlit, open conditions are naturally transient in woods as open areas are invaded by scrub or the canopy closes overhead, increasing the level of shade. Pollination rates are therefore very variable: a study in Hants found that an average of 35% of flowering plants in sunny glades produced at least one seed capsule; under the shade of a high canopy this fell to 16% and in scrub to just 7%. Sword-leaved Helleborine will persist and can flower well under the shade of a high, closed canopy but, in the absence of successful pollination and reproduction, will eventually disappear.

As well as sunlit sites, higher rates of pollination are associated with larger congregations of flowers, and it may be that the spectacle of many helleborines in flower is more attractive to bees. Conversely, at many sites, especially where there are very small populations, virtually no seed is produced.

Self-pollination has not been recorded and is unlikely due to the structure of the flower.

DEVELOPMENT & GROWTH

Forms a relationship with ectomycorrhizal fungi (in Europe basidiomycetes, mostly belonging to the Thelephoraceae), associated with Beech trees in S England, Corsican Pine on Anglesey and very probably Sessile Oak in Wales and W Scotland. A wider range of fungi are found associated with seeds and with adult plants than the much more restricted suite of species associated with the underground seedling stage, which are all crust fungi, genus *Tomentella*. It seems probable that, like Red and White Helleborines, it acquires a substantial proportion of its nutrients via a fungal route - studies in Europe have found that 33% of the orchid's carbon comes from fungi. Plants are able to remain dormant underground for about one year.

No information from wild plants on the length of the period between germination and the first appearance above ground but it may be four or more years. In the laboratory the interval is two and a half years. Plants need to have reached a certain height and minimum leaf area before they can flower but do sometimes bloom in their first year above ground.

Appears to lack any mechanism for vegetative propagation.

STATUS & CONSERVATION

Nationally Scarce and listed as 'Vulnerable'. One of the most threatened British orchids. Has been in decline throughout the 19th and 20th centuries, initially due to woodland losses and the replanting of woodlands with conifers. At a few sites, small populations were given the final death-knell by collectors. The decline has continued in the last 40 years, probably due to the lack of woodland management leading to the disappearance of glades and rides and the development of a denser canopy; management that does take place

and the prospects of such tiny populations surviving are bleak. By 2003 only nine sites in Britain held more than 50 plants, and in Ireland the species was formerly recorded from 15 counties but is now extinct in ten of them.

Effective conservation is complex. Long-term survival depends on adequate levels of flowering and fruiting and the subsequent recruitment of enough new plants to the population. It will flower well once the light levels reach a critical threshold (which is rather lower than the levels produced, for example, by coppicing). Its pollinating bees require higher levels of sunlight, however, and at these higher light levels there is a tendency for rank vegetation and scrub to take over and this will overwhelm the helleborines. 'Nutrient creep' from surrounding agricultural land can also promote the growth of rank vegetation that can smother the orchids.

The most effective management regime probably comprises limited felling to produce a mosaic of woodland and glades, winter grazing to suppress rank vegetation and protection from deer and rabbits when flowering and fruiting. Recent successes give hope for this very special orchid: at the Little Shoulder of Mutton in Hants, 31 plants were found in 1987 after Beech trees and scrub were cleared to increase the size of an existing small patch of chalk grassland. A programme of further scrub control, mowing and protecting individual plants with wire netting resulted in a negligible increase, but dramatic results came when winter grazing was introduced, with a jump to 240 plants by 2004. It now seems to be the 'perfect' site with a sunny, south-facing scrubby edge adjoining chalk grassland that provides nectar and nest sites for pollinating bees.

often involves clear-felling large areas at once. The loss of nearby hedgerows and permanent pastures may have indirectly affected the species too, reducing the populations of suitable pollinating bees. Other threats include heavy browsing by deer, now commoner in Britain than since the time of William the Conqueror.

Sword-leaved Helleborine can now be found at twenty sites each in England and Scotland and six in Wales. The strongholds are in Hants, with ten sites, and Argyll, with six sites. Research in the 1990s showed that 31% of sites supported just one plant and 61% held fewer than ten. Many are likely to be long-lived individuals clinging on at a site no longer suitable for reproduction,

DESCRIPTION

Underground The aerial stems grow from a rhizome that produces long and sparsely branched roots. **Stem** Single, green, with some short hairs on upper part and 2–4 whitish, often green-tipped sheaths at base. **Leaves** 7–20, spaced alternately up stem and arranged in two variably defined opposite rows that may twist around the stem in a spiral; held at 30°–60° above the horizontal; grass green with fairly prominent veins, strap-shaped, keeled, long and rather narrow, upper pointed, lowermost shorter, broader and blunter. **Spike** Rather loose, with 3–15 flowers (sometimes 25, very exceptionally 40), the lower held at 30°–45° above the horizontal; unopened upper buds tend to be nearer vertical. **Bract** Green, narrow and pointed. On the lowest few flowers either very long and leaf-like or much shorter, only 1/2--2/3 length of ovary; bracts always very short at tip of spike. **Ovary** Green, long and slender, prominently twisted and boldly 6-ribbed, with short hairs; the very short stalk is minutely hairy. **Flower** Sepals and petals white, sepals lanceolate, pointed; petals slightly shorter and broader. **Lip** White, short and broad and divided into inner (hypochile) and outer (epichile) halves by a distinct narrowing or 'waist'. Hypochile and base of epichile held parallel to column with their sides curved upwards and inwards to form a deep gutter; epichile spade-shaped with a projecting central lobe or 'tooth'. Base of hypochile washed rich golden-yellow. Towards tip of epichile is an extensive patch of dense papillae with a golden-yellow frosting, and towards the base of this patch 5–6 (rarely 7) parallel longitudinal ridges, also washed golden-yellow; fringe of lip white.
Column Whitish, long, projecting forward (like a boat's figurehead); anther tinged pale-yellowish. **Scent** None.
Subspecies None. **Variation** None.
Hybrids *C.* x *schulzei*, the hybrid with White Helleborine, has been recorded very rarely from Hampshire and West Sussex.

RED HELLEBORINE *Cephalanthera rubra*

IDENTIFICATION

Extremely rare. Confined to single sites in Buckinghamshire, Gloucestershire and Hampshire that between them have held a total of *c.* 50 plants in recent years, with perhaps only ten flowering spikes.

Height 15–60cm. Identification is likely to be academic – if you are fortunate enough to be looking at a Red Helleborine you will have travelled specifically to see it (although visiting the remaining sites is generally not encouraged). It is possible, however, that new sites may be found. One of the most striking and attractive of all British and Irish orchids, the flowers are unmistakable, being a beautiful shade of pink. When fully open the lateral sepals are held horizontally like outstretched arms, while the upper sepal and petals form a rather loose hood. Red Helleborine is, however, shy to flower and can remain in a vegetative state for many years. Even when it flowers, there may be many more non-flowering plants in the vicinity which are hard to spot and thus easily trampled. **Similar Species** Non-flowering plants are difficult to distinguish with certainty from White Helleborine or the *Epipactis* helleborines. **Flowering period** Mid June–mid July (sometimes to late July).

HABITAT

Beechwoods on chalk in the Chilterns and Hampshire, or beechwoods on limestone in the Cotswolds, usually on free-draining slopes. At all three current sites it grows at the boundary of the chalk or limestone and the overlying acidic clay drift. All three sites are probably ancient woodland.

POLLINATION & REPRODUCTION

Pollinated by small solitary bees. The flowers do not produce nectar, and thus deceive their pollinators, which get no reward. It has been suggested that the helleborine flowers mimic various bellflowers *Campanula* spp., plants that do produce nectar in order to attract bees, and that they typically grow in the vicinity of bellflowers. To the bees (whose eyes are not sensitive to the red end of the spectrum) they resemble blue bellflowers. Not only do bees visit the bellflowers to collect nectar, but also male bees search for females around the flowers. Red Helleborine may therefore be able to exploit the bee's sexual urges as well as its foraging behaviour.

In Europe males of the solitary bees *Chelostoma fuliginosum* and *C. distinctum* act as pollinators. Neither occurs in Britain, however, and although the closely related *C. campanularum* does and is a regular visitor to the flowers, it appears to be too small to pollinate the orchids. Not surprisingly, very few flowers set seed in Britain (just one mature capsule recorded in 10 years at the Chiltern site). Hand pollination has resulted in *c.* 50% of flowers producing seed pods, and attempts have been made to germinate seed in the laboratory at Kew, but so far without success (previous analyses of seed from the Chilterns found only *c.* 20% were viable).

Vegetative reproduction may be more important than reproduction by seed.

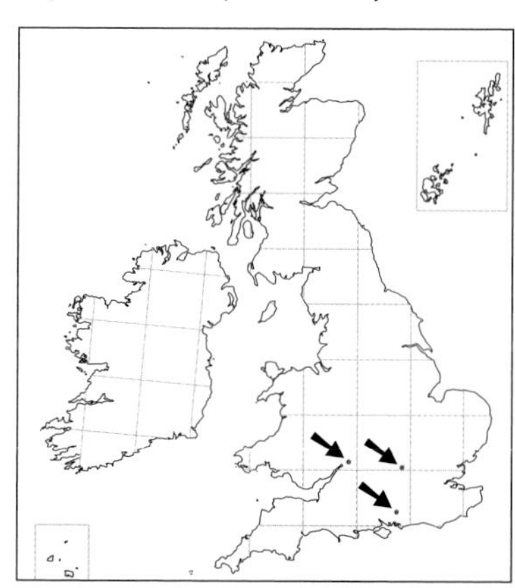

If the central rhizome dies off, the short side roots, densely infected with fungi, can remain alive and produce a bud at the tip that will grow into a new rhizome and eventually produce a new leafy shoot.

DEVELOPMENT & GROWTH

Mature plants acquire *c.* 60% of their nitrogen and *c.* 25% of their carbon via their fungal partner. Red Helleborine has a particular association with ectomycorrhizal fungi and may gain nutrients via these fungi from the roots of nearby trees, especially Beech. But unlike the tree, it cheats the fungus, giving nothing in return.

May become dormant underground, and gaps of up to four years between above-ground appearances have been recorded (although it is possible that shoots did emerge but were quickly grazed off). Indeed, it may be able to persist underground for much longer. A plant appeared in Hampshire in 2003 that had not been recorded in any of the previous 17 years. With 14 flowers it was one of the most robust plants seen at this site, suggesting that it was not a newly emerged seedling.

No information on germination or early development. It has been stated that the first leaves appear about six years after germination, with the first flowers when the plant is ten years old. But, as with all such reports, the actual time scale may be much shorter.

STATUS & CONSERVATION

Nationally Rare and listed as Critically Endangered; WCA Schedule 8. The first British record, in 1797, was from the Cotswolds, and for a long time this area was the stronghold of the species, with records from many of the beechwoods from Nailsworth to Birdlip. Although it was generally scarce, and flowered erratically, it was common enough at times: 'I once saw some fifty or sixty plants together, but only about ten bore spikes of flowers, and somebody cut those before the next morning' (Riddelsdell *et al.* 1948). The number of sites gradually dwindled, however, and there is now just one, where numbers fell in the late 1990s to just three plants. Happily, 42 plants were recorded in 2013, presumably as a result of positive management. In Bucks Red Helleborine was found in the Chilterns in 1955 and a small population is still present. Red Helleborine was recorded in Hants in 1926 and then found in N Hants in 1986; unfortunately, no plants have been recorded at this site since 2008, but it may well still be present, hiding underground. There are old records from W Glos and it probably once occurred in Sussex and possibly also in S Somerset and Kent.

Genetic studies conducted at Kew have shown that the colonies in Bucks, Glos and Hants are genetically distinct from each other and probably represent separate migrations from Continental Europe.

Although a woodland orchid, too much shade will prevent it from flowering regularly or successfully. It has long been known that the finest specimens were to be found in rather open spots within the woods, sometimes amongst tall grass, brambles and other undergrowth alongside paths or in scrubby places and on open banks. In such open situations, however, the growth of scrub can eventually overwhelm the helleborines. It may well be adapted to flower in the gaps caused by treefalls, where a sudden increase in light allows it to flower for a year or two before the canopy closes again, light levels fall and the plant retreats to a dormant state underground until conditions become favourable again. Alternatively, grazing animals in the primeval forests may have kept the development of rank vegetation and scrub at bay for long periods of time. In France and Germany, where Red Helleborine is locally common, it thrives best in situations where it receives a few hours of direct sunlight each day, in glades or along forest roads and tracks.

DESCRIPTION

UNDERGROUND Aerial stems grow singly from a slender, horizontal rhizome. **STEM** Slender, often wavy, with abundant short glandular hairs on upper portion; dusky green, variably washed brownish-purple towards tip. **LEAVES** Dusky grey-green, held at *c.* 45° in two ranks alternately up stem. Most are long, narrow and lanceolate, but lower leaves shorter and blunter. **SPIKE** Rather open, with 2–17 flowers. Ovaries held more-or-less erect: the flowers face upwards and outwards. **BRACT** Very narrow and pointed, lower *c.* 1.5x the length of ovary, upper roughly equal in length; dusky green, with numerous short glandular hairs, especially towards base. **OVARY** Slim, cylindrical, ribbed and twisted, with numerous glandular hairs. Dusky green, becoming brownish-purple at base; ribs variably washed purplish. **FLOWER** Sepals and petals lilac-pink, becoming whiter around base, with numerous short glandular hairs on outer surface. **LIP** Divided into hypochile and epichile by a constriction around the mid-point. Basal portion (hypochile) gutter-shaped, white with fine yellow veins and pale-pink sides that curl up on either side of the purplish column and anther. Epichile flatter (but still concave), arrow-shaped and tapering to a pointed tip that is bent downwards. Whitish with variably pinker edges, a deep lilac-pink tip and 7–9 longitudinal yellow ridges. **SUBSPECIES** None. **VARIATION** None. **HYBRIDS** None.

INTRODUCTION TO HELLEBORINES, GENUS *EPIPACTIS*

Epipactis helleborines are relatively easy to recognise as such. They have upright stems 10–120cm tall and oval leaves with obvious parallel veins. In some species, several spikes can arise from the same rootstock. When it emerges from the soil the stem is bent double and as it grows it continues to 'weep'. Eventually, however, the stem becomes fully upright and the flowers, which remain in bud for a frustratingly long time, begin to open. In many species the flowers are relatively small and drab but they may be very numerous. All the helleborines bloom from mid to late summer, later than many other orchids.

POLLINATION

The *Epipactis* helleborines fall into two groups. Four species are cross-pollinated, with insects (often wasps) transferring pollinia from one flower to another. Four are self-pollinated – their pollinia break-up in the flower and fall onto the stigma below, effecting pollination. In one species, Green-flowered Helleborine, the flower may not even open, self-pollinating in the bud. When identifying helleborines it is important to establish whether it is cross-pollinated or self-pollinated – an accurate diagnosis of this feature is vital to any difficult identification.

GROWTH PATTERN

The aerial stem grows from a rhizome that puts out numerous roots, each of which can live for around three years. The growth pattern of the rhizome is zigzag (sympodial) – the tip of the rhizome grows upwards to form a flower spike that withers and dies off once seed has been set, while the rhizome continues to grow from one or more buds that are formed at the base of the aerial stem. These buds are formed at least a year before the next stem appears above ground but may remain dormant for one or more years, in which case the plant remains underground for a year or more. The probability of a plant appearing above ground and blooming is therefore determined by growing conditions over at least the previous 12 months.

DEVELOPMENT FROM SEED

Poorly known, as the subterranean seedlings are difficult to find. The seeds have a well-developed outer shell or carapace and this slows the uptake of water and delays germination, suggesting that they have a period of dormancy. Seed probably germinates in spring, forming first a protocorm and then a mycorhizome (the earliest and most heavily infected stage in the development of the rhizome). The first roots develop in the autumn and, as the rhizome grows, fungal activity is transferred to the roots so the rhizome itself becomes 'infection-free'. Research has shown that fungi contribute very significantly to the nutritional budgets of helleborines, even as adults, with several having connections via their fungal partners to neighbouring trees.

VEGETATIVE REPRODUCTION

This may take place via several mechanisms. 1. The roots develop buds, which go on to form a secondary rhizome (recorded for Dark-red Helleborine). 2. The rhizome produces two flower spikes that each produce buds and adventitious roots below ground, and these buds go on to produce new rhizomes (recorded for Dune Helleborine). 3. The rhizome branches (as in Marsh Helleborine). In all these examples, if the central 'mother' plant dies off or the rhizome is broken up in some other way, two or more new plants may result. However, vegetative reproduction is not recorded for several of the helleborines and appears to be relatively unimportant for most species.

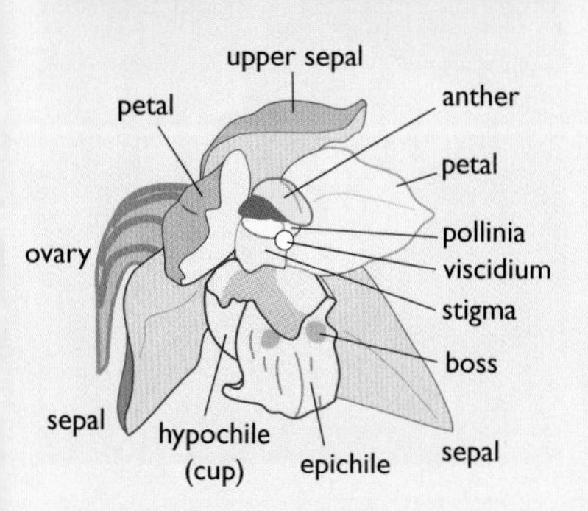

SEPARATING CROSS-POLLINATED AND SELF-POLLINATED FLOWERS

Broad-leaved Helleborine (upper left) is pollinated by wasps, which transfer the pollinia from one flower to another. The pollinia are attached to the wasp by the viscidium, a small, whitish, spherical body about the size of a grain of sugar, which when ruptured releases a botanical 'super-glue'. When the flowers first open the creamy-yellow pollinia sit on top of the column and, although they are covered by the anther cap, can usually be seen poking out like fat yellow sausages. The viscidium is *clearly visible* as a more-or-less spherical white blob. Once a wasp has visited, however, both the pollinia *and* the viscidium will have been *removed*, usually *intact*, leaving no trace. The flower will probably be fertilised soon after and the whole column will start to look tatty and brown. Thus *both pollinia and viscidium are present, or both are absent*. Note that the *uppermost* open flowers on the spike are always the freshest and the best place to look.

Narrow-lipped Helleborine (lower left) is self-pollinated. The pollinia are released by the anther and sit on top of the column. By the time the flower opens they have begun to swell and crumble, fragments falling over the edge onto the stigma below. The pollinia are largely hidden by the anther cap but appear to 'foam-out' from under it. The flower, having been self-fertilised, starts to go-over and the column turns brown. Thus the pollinia are *never removed*, and there is *no viscidium* (occasionally a small viscidium may be present just as the flower opens, but it quickly withers).

FLOWER STRUCTURE

The column is a robust structure that projects from the ovary into the centre of the flower. In *Epipactis* it is often conspicuous. It is topped by the anther or 'anther cap', which is attached to the remainder of the column by a short stalk or narrow flexible hinge. The two pollinia develop side-by-side within the anther.

In *Epipactis* the lip is divided into two parts. The inner section is known as the hypochile; it is cup-shaped and often contains nectar. The outer part, or epichile, is flatter, more or less triangular in shape and often reflexed (bent under) at the tip. At the base of the epichile there are raised areas known as bosses and in some species these are contrastingly coloured. The hypochile and epichile are connected by a narrow strip that is rigid in most species.

Marsh Helleborine

Cross-pollinated. FLOWERS Large and opening widely. Sepals and petals washed dull brownish-purple. Lip usually has *purple veins at the base*, tip *white with a frilled* edge. Uniquely, the inner and outer sections of the lip are connected by a *narrow flexible hinge*. STEM Very hairy, purplish-brown.

Dark-red Helleborine

Cross-pollinated. FLOWERS Often not opening widely. *Reddish-purple with contrasting yellow anthers and pollinia.* Lip with two *large and elaborate wrinkled bosses that merge into a V- or heart-shape.* STEM Dark, densely hairy. LEAVES *Dull, dark green, distinctly folded* and held in two opposite rows.

Narrow-lipped Helleborine

Self-pollinated. FLOWERS Held drooping, but large and opening widely. Colours *clean,* overall pale green, lip whiter, with a hint of pink, *long and pointed, tip not curved down and under.* Cup at base of lip with a dark lining. Base of flower stalk *greenish*. Ovaries *finely hairy.* STEM *Finely hairy.*

Dune Helleborine

Self-pollinated. FLOWERS Small and cup-shaped. Colours dull, but petals and base of lip washed pink. Tip of lip usually curved under, cup at base *with a dark lining*. Base of flower stalk usually tinged purplish. STEM Upper stem finely hairy. NB Inland in N England 'Tyne Helleborine' can be very like Narrow-lipped Helleborine.

Violet Helleborine

Cross-pollinated. **FLOWERS** Large and opening widely. Colours *clean and bright*; sepals and petals pale green, lip whiter, with a hint of pink. **STEM** The large pale flowers *contrast strongly with the dark stem*. **LEAVES** Relatively small and narrow, *dusky green*, usually with a *characteristic purplish wash to the underside*.

Broad-leaved Helleborine

Cross-pollinated. **FLOWERS** Large and opening fairly widely. Colour *very variable;* sepals, petals and lip may be strongly washed pink or pinkish-purple, but may also be pale green with little pink or purple. Base of flower stalk *washed purple*. **STEM** Finely hairy. **LEAVES** Usually rather broad.

Lindisfarne Helleborine

Self-pollinated. **FLOWERS** Small, not opening widely. Colours *dull,* overall greenish-yellow, lip whiter. Cup at base of lip with a dark lining. Base of flower stalk greenish. **STEM** Upper stem finely hairy. NB The only helleborine on Lindisfarne.

Green-flowered Helleborine

Self-pollinated. **FLOWERS** Usually not *opening widely* (sometimes hardly opening at all). Overall pale green, lip whiter, sometimes with a hint of pink. Cup at base of lip *lacks an obvious dark lining*. Base of flower stalk *greenish*. Ovaries *hairless*. **STEM** Upper stem *hairless* (occasionally sparsely hairy).

MARSH HELLEBORINE *Epipactis palustris*

IDENTIFICATION

Very locally common or even abundant. Height 20–45cm (10–80cm). Distinctive. Easily identified by its flowers, which look very much like a *minature hothouse orchid*; they resemble those of other helleborines but are relatively large and bright, and open widely. The lip projects forwards horizontally and usually shows striking *purple veins at the base* and has a large white tip with a distinctive *frilled edge*. Uniquely, the inner and outer sections of the lip are connected by a *narrow flexible hinge* – flicked down with a finger tip, it will bounce back up again. **SIMILAR SPECIES** None. **FLOWERING PERIOD** Late June–early August, very exceptionally to early September.

HABITAT

A wide variety of wet, marshy habitats, but requires neutral to alkaline ground water and relatively short, open, vegetation to thrive. The two most typical habitats are dune slacks, where the ground water is calcium-rich due to the presence of shell fragments in the sand, and spring-fed fens where the ground water is both nutrient-poor and calcareous. Suitable fens may be found in heathland valleys or within more extensive acid bogs, as well as on more obviously chalk or limestone-rich soils. Occasionally also found in meadows that are seasonally flooded with chalky water, but it cannot compete with tall vegetation and such habitats must be regularly mown or grazed for it to survive. Occasionally grows in other habitats, including slumped, clay cliffs, gravel pits and fly-ash pools. Abundant on the floor of an old chalk pit in Norfolk where there is standing water in the winter months, and very occasionally grows in small numbers in 'dry' chalk grassland, especially where quarrying and excavations have left a compacted surface prone to waterlogging.

POLLINATION & REPRODUCTION

Usually insect-pollinated, although the structure of the flower permits self-pollination, especially as the flower ages. Typically *c.* 80% of flowers set seed. The flowers contain a small amount of nectar and are visited by a wide variety of insects. There is debate as to which are the most efficient pollinators, and it probably varies from place to place, but hoverflies, small bees and solitary wasps are undoubtedly important. It is also not clear what role the unusually flexible outer part of the lip plays in pollination. Charles Darwin suggested that it hinges downwards due to the weight of the visiting insect, allowing it to enter the flower without removing the pollinia. Once the insect is within the hypochile, however, the epichile hinges back up to its original position. The insect, as it backs out of the flower, is therefore forced upwards, allowing the pollinia to become attached to its head. Darwin's mechanism is still debated, but the hinge may have the specific function of helping to attach pollinia to an appropriate insect's *thorax*, in which position it is hard for the insect to remove it by grooming.

Vegetative reproduction may occur if the rhizome breaks up into several sections.

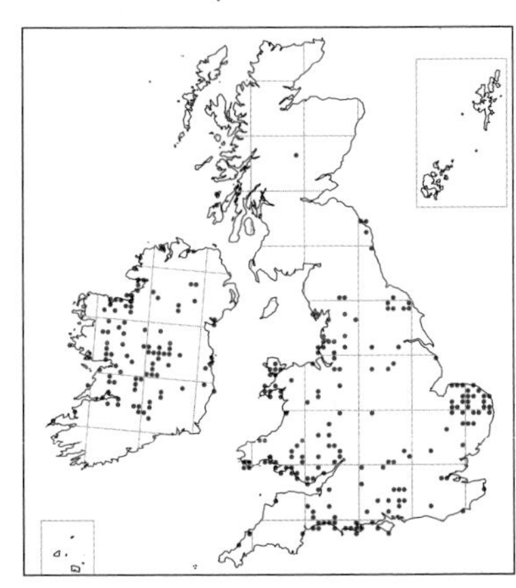

DEVELOPMENT & GROWTH

May acquire *c.* 30% of its nitrogen from its fungal partner but it does not appear to receive any carbohydrates via that route.

Seeds probably germinate in the spring, having been dispersed in late summer and been subject to several months in wet ground before being chilled by winter temperatures (trials in cultivation suggest that the seeds need some after-ripening and that the seed case needs to be partially decomposed before germination will occur).

STATUS & CONSERVATION

Has declined substantially and now gone from 60% of its total historical range in Britain and 39% in Ireland. The decline has affected all areas but perhaps especially those away from the coast. Many of the losses occurred in the 19th century due to the drainage and destruction of marshes and fens, and these factors, together with the subtler effects of water abstraction, continued to cause losses in the 20th century. More recently, eutrophication, that is the enrichment of ground water by fertiliser run-off and the discharge of phosphates in treated sewage, has caused suitable fens to become overgrown with more vigorous vegetation. The abandonment of grazing or mowing compounds this effect and leads to the invasion of fens by scrub and the eventual shading-out of the orchids.

▲ Var. *albiflora*

DESCRIPTION

Underground The aerial stems grow from a relatively slender, well-branched rhizome that creeps horizontally near the surface of the soil. A single plant may have an extensive rhizome and produce several aerial stems (it has been claimed that over 100 flower spikes may grow from the same plant). Roots are produced at many points along the rhizome, both

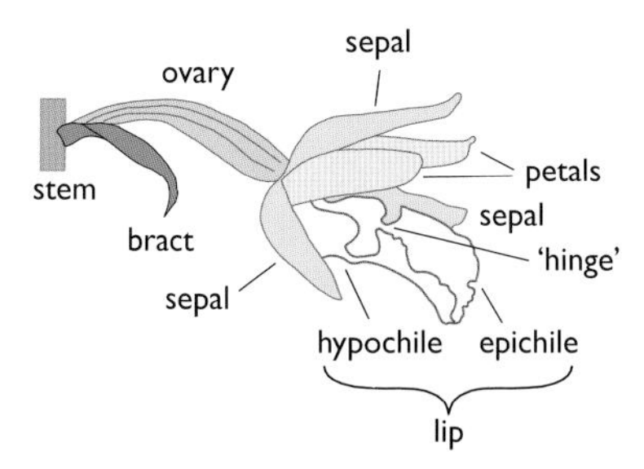

horizontal roots that penetrate the more organic surface layers and vertical roots that often grow deep into the mineral soil. **STEM** Green, flushed brownish-purple, especially towards tip, and prominently hairy. **LEAVES** 4–8, crowded at base of stem, held rather stiffly upwards at *c.* 45° and folded to form a distinct 'keel'; broadly strap-shaped with a pointed tip, papery in texture with 3–5 prominent veins. Upper leaves much smaller, narrower and more bract-like. **SPIKE** Up to 25 flowers, although usually rather fewer, form a loose spike; all face more or less to one side and are initially held horizontally but slowly droop. **OVARY** Brownish-purple, prominently hairy, cylindrical, with 6 fine ribs and tapering to a short, purple stalk. **FLOWER** Sepals sparsely hairy, dull greenish-yellow, flushed, veined and mottled purple, with a whitish border. Upper petals white, veined and washed pinkish-purple towards base. **LIP** Hypochile gutter-shaped, base yellow, variably blotched with reddish nectar-producing swellings, sides white with purple veins. Epichile white, more-or-less circular with the sides turned upwards and strongly frilled or crimped; an irregular boss at the base is edged yellow and bisected by a deep, narrow groove; it also produces small quantities of nectar. **COLUMN** Yellowish-white with well-developed white viscidium, dull yellow anther cap and primrose-yellow pollinia. **SCENT** None. **SUBSPECIES** None. **VARIATION** **Var.** ***ochroleuca*** lacks brown and purple pigments: stem and ovaries green, sepals yellowish-white to pale green, petals and lip white, but interior of hypochile still with purple veins. Scarce, but where found may occur in large numbers. **Var.** ***albiflora*** is similar but lacks purple veins. It is rare. **HYBRIDS** None.

DARK-RED HELLEBORINE

Cephalanthera rubra

IDENTIFICATION

Very local in open, rocky places in the N and W of Britain and Ireland. Height 11.5–60cm (–100cm). Distinctive, the *reddish-purple flowers* have *contrasting yellow anthers and pollinia*, while the lip has two *large and elaborate wrinkled bosses that merge into a V- or heart-shape*. The flowers have a functional viscidium and are cross-pollinated (see p.39) but often do not open widely and may be bell-shaped and slightly drooping. **Similar species** Broad-leaved Helleborine is the only other helleborine to occur in the same rocky habitats, albeit infrequently. It occasionally has rather dark-reddish flowers and, conversely, Dark-red Helleborine may rarely have paler pinkish or greenish-red flowers, similar to those of some Broad-leaved Helleborines. Dark-red Helleborine can always be distinguished by its leaves, which are darker, more markedly folded and held in two opposite rows. It also has rather larger and more wrinkled bosses on the lip and a very hairy ovary. **Flowering Period** Early June –early August, but mostly late June–late July.

HABITAT

Strongly associated with limestone, growing on cliffs, scree slopes, rocky hillsides, in old quarries and the shelter of grykes on limestone pavement. Usually found in the vicinity of bare rock but sometimes also on well-drained grassy slopes with scattered scrub or even in meadows or on road verges. Most sites are open and sunny, but also found in moderate shade in well-wooded limestone pavements, open ash woodland or pine plantations. Indeed, light woodland and woodland edges may be the more typical habitat, as in Europe – most British sites are deforested and heavily grazed and the species has a very fragmented, 'relict' distribution. Most sites are between sea level and 270m, but grows at over 500m in Upper Teesdale and 610m in E Perthshire.

POLLINATION & REPRODUCTION

Insect-pollinated, mostly by bumblebees, although wasps and hoverflies also visit the flowers. Self-compatible, but the flower structure results in generally few or no flowers self-pollinating. Vegetative reproduction may also take place, new plants developing from buds on the roots.

DEVELOPMENT & GROWTH

Isotope studies shown that Dark-red Helleborine acquires *c.* 65% of its nitrogen and 15% of its carbon from fungi. Has an association with ectomycorrhizal fungi and thus, like Coralroot and Birdsnest Orchids, may gain nutrients from the roots of nearby trees via these fungi. These studies, however, took place in Europe, where Dark-red Helleborines are often found in wooded environments. British and Irish plants may grow well away from trees, and the absence of suitable host trees may be a limiting factor in its distribution. No information on the period between germination and flowering.

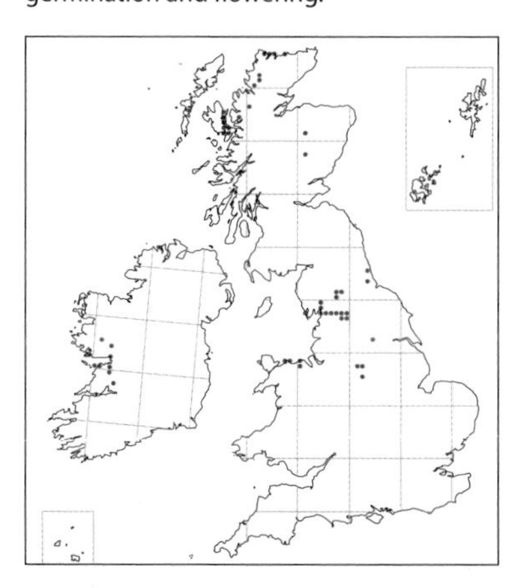

STATUS & CONSERVATION

Nationally Scarce. Very local and absent from many apparently suitable sites within its restricted range. Most populations are small, with a high proportion of non-flowering plants – many that would flower are prevented from doing so by sheep, deer or rabbits. The species has gone from *c.* 30% of its total historic range; some site have been lost due to quarrying, and overgrazing threatens others. Both the British and Irish populations may still be in decline. Bishop Middleton quarry in Co. Durham holds 2,000 or more plants, probably more than all the other British sites put together.

DESCRIPTION

Underground The aerial stem grows from a short, thick, hard rhizome that puts out 40–50 long, slender, widely spreading roots. These sometimes form irregular swellings from which fresh rootlets grow. **Stems** Dull greenish-grey, variably washed purple, especially towards base (sometimes entirely purple), with a dense covering of whitish hairs, particularly on upper part. Stems usually grow singly but sometimes 2–3 may arise from the same rhizome. **Leaves** More or less in two opposite rows, usually with a distinct gap between uppermost leaf and lowest flower. Dark green, variably washed reddish-purple on underside and sometimes purple at base; distinctly longer than wide (the lower elongated-oval, the upper tending to be narrower and more lanceolate), strongly folded and keeled, and held stiffly at *c.* 30° above the horizontal. **Bracts** Green, sometimes washed purple at base; the lower bracts may be a little longer than the flowers

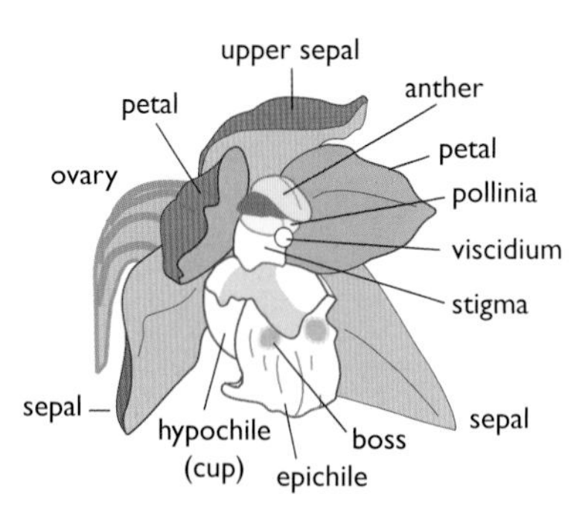

but they become shorter towards tip of spike. **Ovary** Pear-shaped, 6-ribbed, green, variably washed purple (sometimes deep blackish-purple) with abundant short, pale hairs. Flower stalk short, greenish to purplish-black. **Flowers** Sepals with a downy outer surface. **Lip** Hypochile dull green on outer surface of cup, becoming rich purple towards front, interior pale greenish-white mottled purple. Epichile heart-shaped, broader than long, variably turned under at tip. **Scent** Vanilla-like. **Subspecies** None. **Variation** Flowers occasionally paler red or even greenish, especially in sites exposed to full sunlight, but normal plants always have some traces of dark red. Two aberrant varieties: **var. *albiflora*** has white or creamy flowers and has been recorded in W Ross & Cromarty; **var. *lutescens*** has yellowish or buff flowers and has been found in Cumbria and the Burren in Ireland. **Hybrids *E.* x *schmalhausenii***, the hybrid with Broad-leaved Helleborine, has been reported from several areas, notably Cumbria. It is fertile and therefore very difficult to confirm because infertility cannot be used to distinguish potential hybrids from plants that are merely aberrant.

VIOLET HELLEBORINE *Epipactis purpurata*

IDENTIFICATION

Local and uncommon. Height 20–70cm (–90cm). Relatively distinctive. Flowers rather large, opening widely, and overall greenish-white with a pinker lip bearing two pink bosses. The large, pale flowers *contrast strongly* with the dark purplish stem and the *rather small, dark* leaves. Cross-pollinated, the flowers have an obvious and functional viscidium, a useful distinction from Narrow-lipped and Green-flowered Helleborines (see p.39). Flowering stems usually grow singly but multiple stems are fairly common, groups of 6–8 not unusual, and a cluster of 38 has been recorded. **Similar species** Broad-leaved Helleborine is much commoner. It has broader leaves, with the lowest more-or-less wider than long (longer than wide in Violet). Its leaves are also a cleaner and brighter green (duller, more greyish-green in Violet, with a distinctive purplish wash to the underside). Its flowers are smaller and often duller and darker with a purplish wash, with the bosses on the lip usually rougher and browner. Like Violet Helleborine, it is cross-pollinated.

Narrow-lipped Helleborine is also found in beechwoods in S England, but is easily separated by its long, pointed lip and much paler stem, flower stalks and leaves.

Green-flowered Helleborine may also be in flower late in the summer, but is easily separated by its more-or-less hairless, green stem and ovaries, green flower stalks and rather smaller flowers that are usually held drooping and seldom open widely. **Flowering period** Mid July–early September, exceptionally late June–late September; typically peaks in early August.

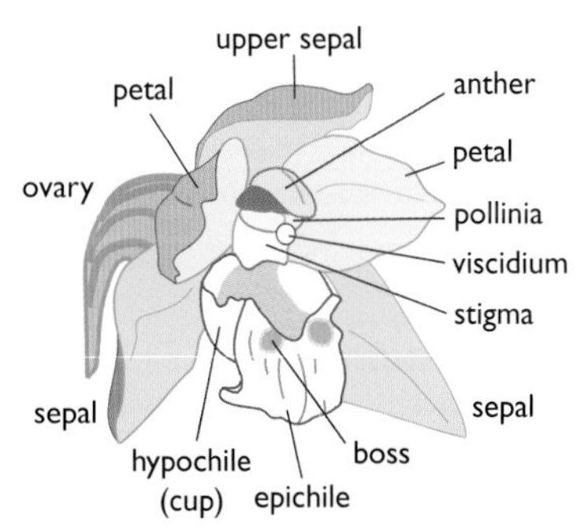

HABITAT

Woodland, favouring beech, hornbeam and oak woods as well as overgrown hazel coppice. Occasionally found in hedgerows that are woodland relicts and in wild gardens. Can tolerate acid soils, but is strongly associated with areas of calcareous bedrock, especially chalk. Its deep root system requires a substantial

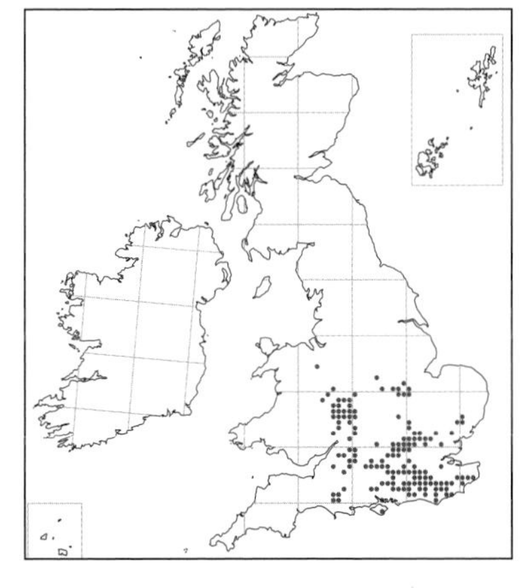

thickness of soil. May grow on sands and gravels but particularly associated with clays, especially the 'clay-with-flints' found on plateaus and hilltops overlying chalk. Within woods, often found in areas of deep shade where little else can grow; it can flourish in much darker situations than Broad-leaved Helleborine.

POLLINATION & REPRODUCTION

Routinely cross-pollinated, often by wasps which are attracted to the nectar. As with Broad-leaved Helleborine, the nectar is reported to have a narcotic effect, with 'drunken' wasps falling to the ground. Pollination is efficient and most or all of the flowers on a spike will set seed. There are no reports of vegetative reproduction.

DEVELOPMENT & GROWTH

It is reported that the roots are fungus-free and therefore the mature plant is phototrophic, depending entirely on photosynthesis rather than fungi for nutrition. However, given the dense shade in which it grows, it seems much more likely that fungi contribute a large part of its nutritional budget; the rare var. *rosea* lacks chlorophyll but nevertheless is able to flower and fruit successfully and must depend entirely on fungi.

Violet Helleborine is long-lived and only appears above ground when mature enough to flower. Immature, non-flowering plants are very rarely seen. A single-stemmed plant may be 30 years old, and it has been suggested that large, multi-stemmed plants are probably hundreds of years old. There is no information on the duration of the period between germination and flowering.

STATUS & CONSERVATION

Has declined steadily over the last 150 years and has vanished from the edge of its range as ancient woodlands have been destroyed or replanted with conifers, although most sites should now be safe from this particular threat. But, favouring the dense shade of closed-canopy woodland, this is perhaps the only species of orchid to have benefited from the abandonment of coppicing in many British woods during the 20th century. Conversely, the great storms of 1987 and 1990 devastated many woods, opening up the canopy and leading to a great reduction in the numbers of Violet Helleborines in affected areas. Another threat is deer, and whole populations of orchids can be grazed off in some woods. Can be very persistent, however, and has even been recorded pushing its way through newly laid tarmac.

DESCRIPTION

Underground Grows from a rhizome lying more-or-less vertically in the soil with up to 50 fleshy roots, each up to 70cm long (exceptionally 120cm), growing vertically downwards. **Stem** Greyish-green, variably but often heavily washed purple, with dense, short, grey hairs on upper part of stem and 1–3 small, purplish-brown sheathing scales at base (uppermost scale often tipped green). **Leaves** Well-spaced up stem and arranged spirally (sometimes in two opposite rows), 4–14, relatively small, more-or-less oval, tapering to a point, usually rather more than 2x as long as wide; upper leaves narrower and more bract-like, lowest short and cowl-like. Leaf posture variable – may be held horizontally with tips slightly drooping or at *c.* 30° above the horizontal. Rather dull, cold, greyish-green, may be washed purple towards tips; undersides have diagnostic purple wash, leaf sheaths also frequently tinged purple. **Spike** Slightly to moderately one-sided, usually 7–40 flowers, but over 100 on some well-grown plants. **Bract** Green, variably washed purple; narrow, lanceolate, held roughly horizontally and longer than flower in lower part of spike, becoming shorter higher up. **Ovary** Green, with six prominent ribs, which may be washed purple, sparsely hairy. Flower stalk purplish, variably twisted. **Flower** Sepals triangular-oval, rather large, pale green, becoming paler towards edges with prominent green midrib on outer surface. Petals smaller, whitish, slightly greener towards centre, with fine green midrib. **Lip** Hypochile translucent-whitish, slightly greener towards base of cup, interior variably washed pale purplish-rose to pale brown or pale greenish (this colour shines through to outside). Epichile short, triangular or heart-shaped, tip strongly folded downwards; whitish with two prominent, smoothly pleated, pink bosses at base. Column whitish; large, conspicuous anther cap very pale yellowish-white with narrow brown stripes at sides, viscidium whitish, pollinia pale yellow. **Scent** Faintly scented. **Subspecies** None. **Variation** May sometimes have variegated leaves, and some may have the leaves very extensively streaked violet. **Var. *rosea*** is rare but stunning. It lacks chlorophyll, the entire plant is rosy pink with whitish flowers. **Hybrids *E. x schulzei,*** the hybrid with Broad-leaved Helleborine, has been recorded widely but is fertile and thus very difficult to confirm (see p.19).

BROAD-LEAVED HELLEBORINE

Epipactis helleborine

IDENTIFICATION

The commonest and most widespread helleborine. Height 25–80cm (10–120cm). Very variable, from a tall, robust, leafy plant to small and weedy with just 1–2 flowers. The flowers can be almost entirely green or almost completely purple but are usually a *mixture of pale green, pink and purple*. The outer part of the lip is heart-shaped, broader than long and usually *turned under at the tip*; the two bosses at its base are usually brownish and rough or wrinkled but can be smooth and pink. The base of the flower stalk is *washed purple*. The flowers are cross-pollinated – when freshly opened, they have an *obvious and functional white viscidium*, an important distinction from Narrow-lipped, Dune and Green-flowered Helleborines (see also p.39). Typically, plants have several relatively large leaves, usually obviously veined, broad, especially the lowest, which may be about as wide as long, and placed all around the stem (but sometimes in two opposite ranks). The upper stem and ovaries are *hairy* (the ovaries sometimes sparsely so). **SIMILAR SPECIES** Narrow-lipped Helleborine has green flowers, often with a delicate pink wash to the petals and lip, but not the sepals. The tip of the lip is narrow and pointed, and held pointing outwards. Its leaves are a paler and more yellowish-green and usually held in two opposite ranks. In combination these features should be distinctive, but the tip of the lip of Broad-leaved Helleborine may not always be reflexed, especially when the flower has just opened. In Narrow-lipped Helleborine, however, the outer part of the lip (epichile) is always longer than broad, the base of the flower stalk is greenish-yellow and the flowers are self-pollinated and lack an effective viscidium.

Dune Helleborine can be separated by its more yellowish-green, 2-ranked leaves, held rigidly at *c.* 45° above the horizontal. Its flowers are smaller, duller and do not open as widely. The petals and base of the lip are variably washed pink, but it never shows pink or purple tones to the sepals or a strong purple wash to the lip. At inland localities a variant of Dune Helleborine, known as 'Tyne Helleborine', is found. This is very like Narrow-lipped Helleborine and can be distinguished from Broad-leaved by the forward-pointing tip to its lip and yellowish-green base to the flower stalk. Dune Helleborine is also normally self-pollinated.

Green-flowered Helleborine typically has green flowers that are held drooping and do not fully open. Sometimes, however, its flowers may be held more horizontally and open widely although they are still predominantly green, with any pink tones restricted to a delicate wash on the lip. Its upper stem and ovaries are hairless or with just a few, sparse hairs on the stem, the base of its flower stalk is green and it is self-pollinating. See also Dark Red Helleborine (p. 46) and Violet Helleborine (p. 50). **FLOWERING PERIOD** Early July–early September, mostly mid July–mid August.

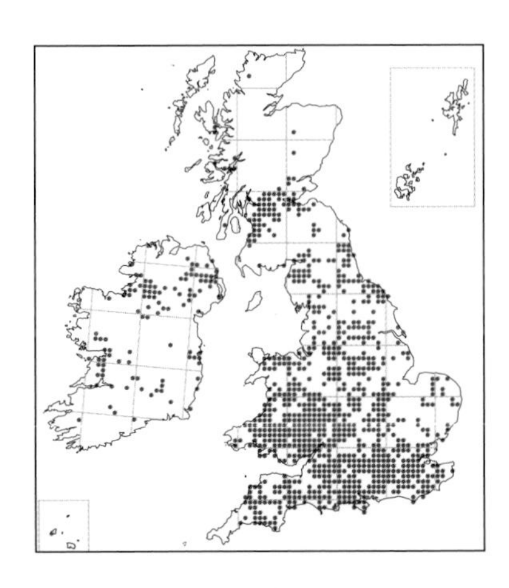

HABITAT

Primarily deciduous woodland, especially better-lit 'edge' habitats along roadsides, paths and rides, and in glades, but will grow in deep shade. Like many helleborines has an affinity for Beech trees. Can also be found in suitable shady conditions in scrub and along well-grown hedges and disused railways. Will grow in the open, on limestone pavements, cliffs, scree and grassland, but only in the cooler and damper conditions of the N and W. In Ireland and S Wales found in dune slacks. An opportunist, it can colonise newly available habitats, such as birch woodland on spoil heaps, willow and alder carr and conifer plantations. In Glasgow and a few other cities in S Scotland and NE England, has moved into mature gardens, parks, cemeteries, golf courses, playing fields, rubbish tips, roadsides and railway embankments – said to be commoner in Glasgow than anywhere else in Britain. Comparatively tolerant of soil pH and will grow in slightly acid conditions, but usually commonest on calcareous soils.

POLLINATION & REPRODUCTION

Pollinated by wasps, especially long-headed species of the genus *Dolichovespula*. Wasps can easily reach the nectar in the hypochile and in the process they rupture the viscidium and the pollinia are stuck to their heads. Other insects, including short-headed wasps, bees, hoverflies and beetles, may visit the flowers but are the wrong size or shape to act as efficient pollinators. Fermentation of the nectar in the flower may produce ethanol and this can have a narcotic effect on visiting wasps, which become slow and sluggish and may even fall to the ground 'drunk'. Plants are self-compatible and frequently pollinated by wasps carrying pollinia from flowers of the same spike, but self-pollination either does not occur or only takes place rarely when small insects carry pieces of pollinia onto the stigma below. It has been reported that in drought conditions the flowers shrivel without opening and may self-pollinate in the bud. Seed-set is usually good and almost all flowers may produce ripe capsules, each containing up to 3,000 seeds.

DEVELOPMENT & GROWTH

The degree of fungal infection of the roots is reported to vary, from high in plants growing in humus-rich soils to negligible in mineral soils, but at least some helleborines are heavily dependent on their fungal partner, acquiring *c.* 60% of their nitrogen and *c.* 14% of their carbon via fungi. Unexpectedly, Broad-leaved Helleborine has an association with ectomycorrhizal fungi and thus, like Coralroot and Birdsnest Orchid, may gain nutrients from the roots of nearby trees via the fungi (see Introduction). Probably not very fussy, however, about its fungal 'partners', but may have a preference for Ascomycetes.

Broad-leaved Helleborine, presumably supplied by its

fungal partner, may spend a significant proportion of the time underground. Plants may flower and then spend one or, rather less often, two or even three years dormant before appearing again. In a study in America, 25%-50% of the population appeared above ground each year and around a third of these flowered. Very few plants flowered every year, however, although annual flowering may be more frequent where the soil is reasonably moist. The interval between germination and flowering can be as little as 18 months, although periods of 8–9 years, including several years above ground but non-flowering, are also quoted.

STATUS & CONSERVATION

Overall distribution stable but there have been significant declines, especially in the Midlands and N England, where the distribution is now rather fragmented, and more recently in the Home Counties. The clearance or 'coniferisation' of woodland, increase in dense shade due to a lack of woodland management, ground disturbance by machinery and horses, and grazing by deer may all have contributed, and losses appear to be ongoing.

DESCRIPTION

UNDERGROUND The aerial stem grows from a small, woody rhizome that sends numerous cord-like roots deep into the soil. **STEM** Pale green, often washed purplish towards base, with short, whitish hairs on upper portion and 2–3 leafless sheaths at extreme base. Usually grows singly but 2–3 stems together are quite frequent and five or more are occasional. Non-flowering stems frequent.
LEAVES Dull, mid–dark green (lacking yellow tones), sometimes washed purple; 4–10, more-or-less spirally arranged around stem, oval to oval-lanceolate, less than 2x as long as wide; lowest leaf sometimes almost rounded, but becoming narrower and more bract-like towards flower spike. **SPIKE** Roughly one-sided,

with up to 60 flowers, rarely to100; from very dense to rather lax. **Bract** Dull green, lanceolate, pointed; lowest significantly longer than flower but becoming shorter, roughly length of ovary, towards tip of

spike. **OVARY** Green, hairless or with a few short hairs, boldly 6-ribbed and rather pear-shaped, tapering to a long, twisted stalk that is washed purple at the base. **FLOWER** Very variable in colour but usually with some purple tones; opening widely. Sepals oval, tapering to a point, outer surface green, variably mottled dull purple or pink (sometimes none) with a prominent green midrib; inner face pale green to dull, dusty purple. Petals similar but slightly shorter and less tapering, tending to be paler, 'cleaner' and often pinker – from pale, dusty pink to purple – often whiter towards centre, with green midrib on outer face. **LIP** Hypochile pale greenish-white, variably washed pink or purple, interior of cup purple to mid brown, glistening with nectar. Epichile dull greenish-white washed pink, to pale pink or dull purple; heart-shaped, broader than long, major part strongly curled down and under, with two bosses at base, usually purplish-brown and wrinkled, but may be green or pink, and sometimes smooth. **COLUMN** Greenish-white, anther cap dull pale yellow with brown stripes at sides, pollinia creamy-yellow. Viscidium white, obvious and functional. **SCENT** None. **SUBSPECIES** None. **VARIATION Var. *youngiana* 'Young's Helleborine'** was described as a new species, *E. youngiana*, in 1982 from plants found in Northumberland, and subsequently identified in Scotland and Yorks. It was reported to differ from Broad-leaved Helleborine in having leaves more-or-less two-ranked and often paler, more yellowish-green. Ovary sparsely hairy to hairless; flowers relatively large, clean and bright; viscidium small and disappearing rapidly, the pollinia disintegrating onto the stigma, causing self-pollination. Column with a long rostellum that, together with two pointed bosses at base of stigma, form a distinctive 'three-horned' shape.

Some botanists were, however, always sceptical about the distinctiveness of

◄ Var. *youngiana*

'Young's Helleborine' and genetic studies have shown that it does not exist as a distinct entity. At each site investigated, the 'Young's Helleborines' were genetically closer to the local Broad-leaved Helleborines than they were to 'Young's Helleborines' at other sites. Also, rather than being self-pollinated, they had a high level of genetic diversity and a population genetic structure indicating that they were cross-pollinating. Plants matching the description of Young's Helleborine in terms of their leaves, ovaries and flowers, found at its classic sites (e.g. Settlingstones in Northumberland) almost all have a large, functional viscidium and are presumably cross-pollinated, supporting the genetic studies. Given the wide variation in Broad-leaved, it is hard to justify even the status of variety for Young's.

Var. *neerlandica* 'Dutch Helleborine' Overall deep green and rather short (15–40cm), with short, stiff, rounded leaves held more-or-less erect and grouped at base of stem, which they closely sheathe. Leaves with a border of tiny teeth that are irregular and fused at the base (use 20x hand-lens; typical Broad-leaved Helleborines have more regular teeth). Spike dense, flowers dull purplish-pink, bell-shaped, not opening widely. Found on North Sea coasts from NE France to Denmark and on the Baltic coast of N Germany, with similar plants in dune slacks in S Wales identified as this form. Although treated as a distinct species, *E. neerlandica*, by some Continental authors, genetic studies reveal little difference between it and typical Broad-leaved Helleborines.

Var. *albifolia* lacks chlorophyll and is pale pink or straw-coloured with white or rosy flowers. Very rare. **Var. *viridiflora*** lacks anthocyanins and has pale green flowers with a greenish-white lip and shows no trace of red or purple. Rare. **Var. *purpurea*** has especially dark purple or reddish-violet flowers. Rare. **HYBRIDS *E. x schulzei***, the hybrid with Violet Helleborine, and ***E. x schmalhausenii***, the hybrid with Dark-red Helleborine, have been reported from several areas, but are fertile and thus very difficult to confirm.

Var. *neerlandica* ➤

NARROW-LIPPED HELLEBORINE

Epipactis leptochila

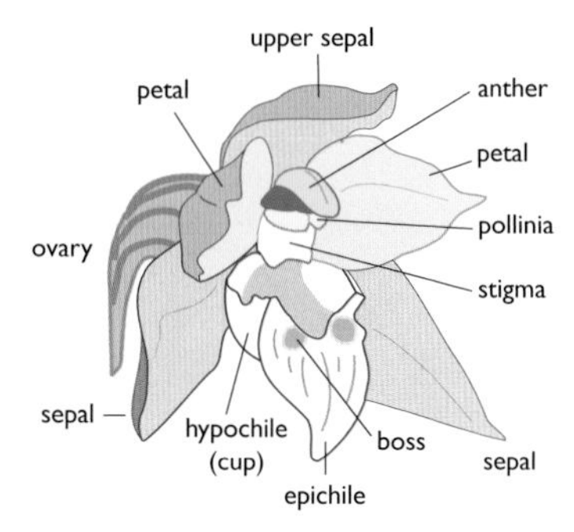

IDENTIFICATION

Scarce and very local. Height 30–60cm (15–75cm). Almost uniformly green, with relatively large, clean, pale green flowers with purplish-pink confined to a delicate wash at the base of the lip and a variably obvious tinge on the petals. Flowers held drooped but usually fairly widely open (occasionally some may not open fully, especially towards top of spike). The tip of the lip *projects forward* rather than being turned under as in most other helleborines and the lip is thus *long and pointed*. Upper stem and ovaries *hairy* (use a 10x hand-lens), base of flower stalk *greenish-yellow*. Leaves usually carried in two opposite ranks. The flowers are self-pollinated and *lack a viscidium* – the pollinia crumble and fall piecemeal onto the stigma – a useful distinction from Broad-leaved and Violet Helleborines (see p.39). Rarely, when freshly open, the flowers of Narrow-lipped Helleborine can be temporarily cross-pollinating, with a functional viscidium and cohesive pollinia; such plants must be identified as Narrow-lipped Helleborine with great caution. **Similar species** Broad-leaved Helleborine can have largely green flowers but they are usually extensively washed with dull pink or purple and the tip of the lip is almost always turned down and backwards to give the lip a very short, blunt-ended appearance. Rarely the lip is not reflexed, but the heart-shaped epichile is always broader than long. The base of the flower stalk is tinged purple and, in most Broad-leaved Helleborines, the leaves are a darker and duller green and are usually carried spirally around the stem. Violet Helleborine has flowers closer in coloration to Narrow-lipped, but they are held more erect and face outwards, giving the spike a different character. The flower stalk is purple and the leaves are a far duller greyish-green and often have a faint purple wash below. Green-flowered Helleborine is self-pollinated, like Narrow-lipped Helleborine, and can be rather similar (especially var. *vectensis* of Green-flowered), with drooping, bell-like green flowers and a green flower stalk. However, it has hairless, or almost hairless, upper stem and ovaries, and the interior of the cup at the base of the lip is pale greenish. In many cases, the flowers of Green-flowered Helleborine are very distinctive as they often hang down almost vertically and in some populations hardly open at all. **Flowering period** Rather short. The second week in July–mid August (rarely from late June) but mostly in the last half of July.

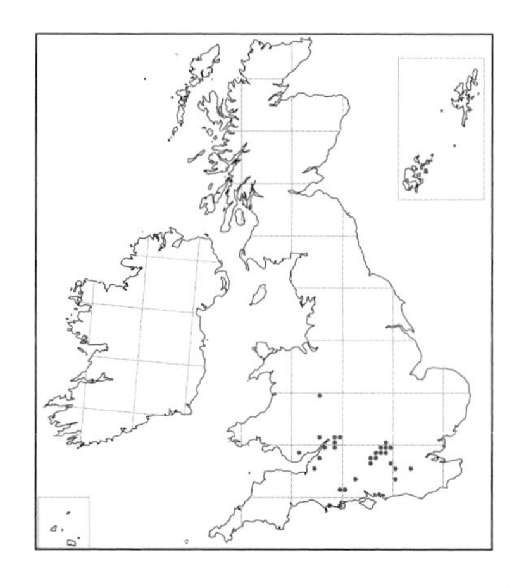

HABITAT

A woodland orchid, usually found in ancient woodland on soils derived from chalk or limestone, especially on steeper slopes where the soil is very thin. Its classic habitat is a beechwood on chalk, but occasionally found under a variety of other deciduous trees, including overgrown ash-hazel coppice. Whatever the type of woodland, typically found in areas of deep shade where the ground cover is sparse or absent, and intolerant of direct sunlight.

POLLINATION & REPRODUCTION

Usually self-pollinated. Seed-set is good, each capsule holding 1,000–2,000 seeds. Very occasionally the flowers have a functional viscidium and insects can carry away pollinia, allowing cross-pollination.

DEVELOPMENT & GROWTH

Little-known. The dense shade of its habitat suggests that Narrow-lipped Helleborine must be able to acquire a large proportion of its nutrients from fungi. Plants lacking chlorophyll, with a yellowish stem and flowers and pale greenish-yellow leaves, have been recorded rarely, and these must be dependent on fungi.

STATUS & CONSERVATION

Nationally Scarce and listed as Data Deficient. Highly localised, the strongholds are the Cotswolds in Glos and the Chilterns in Berks, Oxon and Bucks. Rare away from these areas with just a few widely scattered populations.

Recorded from just 29 10km squares from 1987–1999, it has vanished from 50% of its historical range, with many of the losses being comparatively recent. Direct threats include the destruction of woodland and the conversion of suitable woods to conifer plantations, although this has largely ceased. The loss of suitable shaded woodland due to the opening up of the canopy by severe gales, such as the great storms of 1987 and 1990, is a subtler threat. The compaction of woodland soils by the use of heavy machinery in forestry operations, horse riding, mountain biking and the spread of wild boar are other potential hazards. However, the most obvious issue is the damage to flowering plants caused by rising numbers of deer.

Initially confused with Broad-leaved Helleborine, it was not until 1921 that Narrow-leaved Helleborine was described as a distinct species. Confusion then arose again in the mid 1970s when plants resembling Narrow-lipped Helleborine were found inland in N England (the so-called 'Tyne Helleborine', see p.64). In some places they were found together with Dune Helleborines and intermediates were reported, leading to the conclusion that Narrow-lipped and Dune Helleborines must be the same species. However, genetic studies have shown that all the plants in N England are Dune Helleborines, and that they are genetically distinct from the true Narrow-lipped Helleborine, which is once again known to be restricted to its classic habitat, the southern beechwoods.

DESCRIPTION

Underground The rootstock lies rather deeply in the soil, with numerous roots. **Stem** Green, upper part hairy. Usually grows singly but sometimes two together and rarely up to 5–6 may arise from a single rootstock. **Leaves** 3–7, fresh green, arranged in two opposite rows, although sometimes not obviously so. Elliptical, mostly more than 2x as long as wide, uppermost long, narrow and grading into the lowest bracts. Rather 'floppy' and held more-or-less horizontally (but uppermost, bract-like leaves are pendant). **Spike** Usually rather lax, with 4–35 flowers, often facing to one side. Initially held horizontally, they droop to a variable extent as they age. **Bract** Pale green and lanceolate; low on spike bracts very long, projecting well beyond flower and often hanging downwards, but shorter towards tip. **Ovary** Pale green, variably hairy, prominently 6-ribbed but not twisted; flower stalk greenish-yellow. **Flower** Sepals oval-triangular, elongated into pointed tips, pale green on outer surface, slightly paler and more whitish-green on inner. Petals slightly smaller and paler, more greenish-white, becoming even paler towards edges but greener towards prominent midrib, variably flushed pale pink with faint pink veins. **Lip** Hypochile whitish-green on exterior of cup, variably flushed purplish-pink at sides, similarly pale on inside but with wine-red or chocolate-brown rear wall; contains nectar. Epichile whitish-green, sometimes delicately flushed pink, longer than wide, pointed tip projecting outwards with two relatively small, smooth bosses at base, each washed purple or purplish-pink; they flank a longitudinal central groove connecting hypochile and epichile. **Column** Greenish-white. Pollinia creamy-yellow, anther cap pale greenish-yellow with narrow cream and broad chocolate-brown stripe on either side; attached to column at rear by projecting spur or stalk (in profile looks a little like the 'leaping jaguar' motif of the famous sports car, a distinction from Broad-leaved Helleborine, in which anther is unstalked). Rostellum reduced in size (less than half as long as anther) and viscidium, although present in bud, almost always withered by the time the flower opens. **Subspecies** None. **Variation Var. *cleistogama*** was described from plants found in Glos on the steep W escarpment of the Cotswolds near Wotton-under-Edge, but this population is apparently extinct. It had pendulous bracts, flowers that did not open and, within the flower bud, a greener lip. (NB The upper flowers of some normal plants may never open.) **Hybrids** None.

DUNE HELLEBORINE *Epipactis dunensis*

IDENTIFICATION

Very local, but can be locally abundant. Occurs on Anglesey, in N England and perhaps also S Scotland, mostly on sand dunes on the W coast but also at an increasing number of sites inland. Height: 20-40cm (–50cm). Dune populations have *relatively small, dull, cup-shaped* flowers that do not open widely. Overall yellowish-green, the petals and base of the lip are washed pink. The outer part of the lip is heart-shaped, usually broader than long, with the tip turning under to a variable extent as the flower ages. The base of the flower stalk has a *violet tinge* and the upper stem is distinctly downy. The yellowish-green leaves are held *rather stiffly in two ranks at c. 45° above the horizontal.* The form found at some inland sites in N England, known as 'Tyne Helleborine', has a cleaner and whiter lip (usually lacking pink tones) and a *greenish* base to the flower stalk: in some populations the leaves are also floppier and the outer part of the lip is *longer than broad* and *not folded downwards*. In all populations, usually *self-pollinated* (see p.39). **SIMILAR SPECIES** Broad-leaved Helleborine typically has broader, darker and greener leaves, arranged all around the stem and not held rigidly erect. It has larger and more widely opening flowers, often with pink or purple tones on the sepals and a distinct purple tinge to the lip. Tyne Helleborines can additionally be separated from Broad-leaved by the greenish base to their flower stalks (washed violet in Broad-leaved) and, in some cases, the long, pointed tip to their lip, which is not turned down and under. Broad-leaved Helleborine is very variable, however, and in N England and S Scotland some have more yellowish-green, two-ranked leaves, although they still have large, widely opening flowers (so-called 'Young's Helleborine', see p.58). In tricky cases check the reproductive structures within the flower: Broad-leaved Helleborine is cross-pollinated and Dune Helleborine usually self-pollinated, only very occasionally showing a functional viscidium (see p.39).

Green-flowered Helleborine is, like Dune Helleborine, self-pollinating and potentially more confusing. However, its upper stem and ovaries are either hairless or have a few sparse hairs, its leaves a clean apple green and often short and rounded, and its flowers hang more-or-less downwards, both in bud and when open. A feature to check in difficult cases is the fringe on the edge of the leaves (use a 20x hand-lens). In Green-flowered Helleborine the tiny, transparent, whitish teeth are arranged into irregular groups separated by gaps, but in Dune Helleborine there is an even fringe of minute teeth.

Narrow-lipped Helleborine can be very similar to some Tyne Helleborines. Indeed, for almost 30 years after its discovery, Tyne Helleborine was thought to be Narrow-lipped Helleborine. Fortunately, separation of the two is academic, as Narrow-lipped is confined to S England, with a few colonies on the Welsh Marches and in S Wales. The sepals and petals of Narrow-lipped Helleborine average larger and slightly

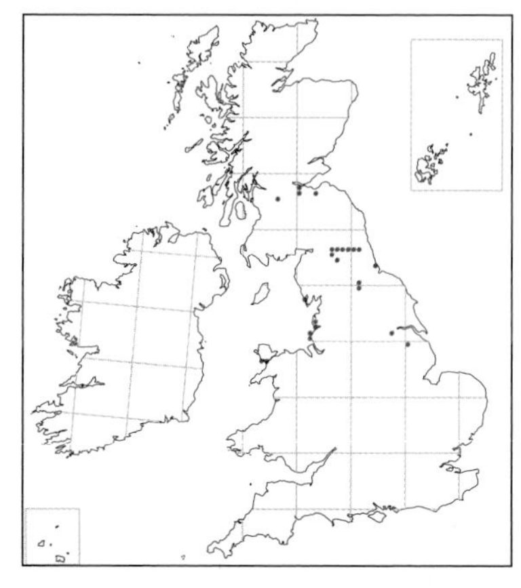

broader, its lip is a little more drawn-out into a point, with the bosses at the base washed pink and the slot between them (the 'keyhole') rather broader. **FLOWERING PERIOD** Late June–mid August, usually peaking in the second week of July in dune populations, a little later in adjacent conifer plantations and inland. Flowers short-lived and can be badly affected by drought.

HABITAT

The classic habitats is coastal sand dunes, where it avoids both bare sandhills and the wet bottoms of the intervening depressions or 'slacks', preferring slightly raised areas within the slacks or around their margins, often growing amongst low-growing Creeping Willows. On the Sefton coast dune-scrub (poplar, birch and Sea Buckthorn) is utilised as much as open slacks, but *c.* 75% of plants are found in pine plantations on the dunes (and pines hold the majority of plants on Anglesey).

Typical 'dune' plants are also found at inland sites, but some inland localities have a distinct form, known as Tyne Helleborine. Inland the habitats is usually light, regenerating woodland, especially birch or willow, in many cases where the ground is kept relatively open by the presence of heavy metals (zinc or lead), mine waste or the clinker of old railway tracks. In Scotland Tyne Helleborine is found on the wooded slopes of old 'pit bings' (the shale-rich spoil heaps produced by coal mining).

It seems likely that Dune Helleborine has spread inland from its semi-natural coastal habitats – all the inland sites are of recent origin and man-made. And, although seemingly very different, adaptations which allow the species to grow in dune slacks prone to the stresses of summer drought and salt spray may be similar to those needed in habitats stressed by the presence of toxic metals.

◀ Tyne Helleborine

POLLINATION & REPRODUCTION

Self-pollinated. Even in bud, fragments of the pollinia may fall onto the stigma and effect pollination. However, in some plants a functioning viscidium may persist until the flower has opened, allowing the pollinia to be removed by insects and cross-pollination to take place. Indeed, even when no viscidium is present, plants may be cross-pollinated by wasps or other insects carrying away the pollinia which stick to their bodies regardless. In one case, *c.* 2/3 of the flowers in a population of Tyne Helleborine had their pollinia removed.

Vegetative reproduction may also occur, the rhizome producing two flower spikes which in turn produce buds and roots.

DEVELOPMENT & GROWTH

No information.

STATUS & CONSERVATION

Nationally Rare and listed as Data Deficient. Endemic to Britain. Recorded from just 26 10km squares in the *New Atlas*. Dune populations are locally common at a few sites on Anglesey, the coasts of Merseyside and Lancashire and by the Duddon Estuary at Sandscale Haws in Cumbria. Here they are sensitive to the level of grazing, especially by rabbits: if too intense the spikes are all nipped off and the plants fail to set seed. Conversely, if grazing is too light, scrub invades the habitat and may eventually shade the orchid out.

Scattered populations of typical dune-type plants have also been recorded at inland sites in Denbighshire (a very large population at Alyn Waters Country Park), Lincs, SE Yorks, Co. Durham, Cumbria, and in S Scotland in Lanarkshire, Midlothian (1994) and West Lothian (but see below). Tyne Helleborine occurs inland in Northumberland, Cumbria, Co. Durham and NW Yorks. Most sites were discovered in the 1960s and 1970s and when found some were large, with over 1,000 flowering plants. All are in young, woodland, however, and as this matures

▲ A typical dune-slack plant, with the tip of the lip turned under.

▲ Tyne Helleborine with a plainer, greener flower and, in this plant, the tip of the lip projecting forwards.

the accumulation of humus seems to buffer the effects of the toxic metals and the ground cover increases; such changes may eventually lead to the disappearance of the helleborines.

Research, based in part on DNA studies, has concluded that all the plants in N England, both 'Dune' and 'Tyne', are closely related and probably each other's closest relatives, with the same ancestor, and have only recently separated into two slightly different entities. These studies also make it clear that they are not the same species as Narrow-lipped Helleborine of S England.

DESCRIPTION

UNDERGROUND Grows from a deeply buried short, slender, woody rhizome that bears up to ten thin, wiry roots. **STEM** Pale green, tinged violet towards base and downy towards tip, with fine, pale hairs. Usually single, but sometimes 2–3 grow from the same rhizome. **LEAVES** 3–10, in two opposite rows up the stem, held stiffly at *c.* 45°; oval-lanceolate, mostly more than 2x as long as wide and relatively narrower towards flower spike, but lowest very short, broad and rounded, forming a cowl-shaped sheath; yellowish-green, deeply veined, margins with fine, regular, whitish teeth (cilia) 0.03–0.06mm wide. Leaves often damaged by wind or drought at flowering time. **SPIKE** Rather lax, with 6–20 flowers, sometimes as many as 35, set to one side of stem. Flowers initially held horizontally but often droop and become pendant as they go over. (At Scottish sites the flowers characteristically held drooping, but this could be evidence of hybridisation.) **BRACT** Strap-shaped, pointed; lower bracts slightly longer than flowers, but becoming shorter towards tip of spike. **OVARY** Green, hairy, 6-ribbed (but not twisted) and pear-shaped, tapering into a short stalk that is usually washed violet at the base. **FLOWER** Sepals yellowish-green, oval-triangular, relatively short and blunt with prominent midrib on outer surface. Petals similar in shape but slightly smaller and very pale green, often washed pink. **LIP** Interior of basal cup (hypochile) reddish-brown to dark brown, containing nectar; exterior whitish, variably flushed pink (the colour shining through from inside). Epichile whitish, sometimes with pink flush in centre, with greener tip; heart-shaped, usually broader than long, tip sometimes folded under, especially as flower matures; often the entire epichile is bent downwards near the join with the hypochile. At base of epichile two smooth, pink or greenish bosses frame a distinct notch running between the epichile and hypochile. **COLUMN** Anther yellowish-green, variably stalked. Pollinia crumbling, whitish. Viscidium, although present in the bud, usually disappears as flowers open. **SUBSPECIES** Two forms have been named as distinct subspecies, but the status of Tyne Helleborine is controversial and it is not recognised by some authorities. ***E. d. dunensis*** Coastal dune systems and some inland sites; see Description. ***E. d. tynensis*** **Tyne Helleborine** Inland sites in N England in the catchment of the River Tyne and in Cumbria. Leaves often rather floppy and not held as rigidly erect. Leaf margins with even finer teeth (0.01–0.05mm wide, sometimes imperceptible). Ovary with a yellowish-green stalk, flower (especially lip) usually lacking pink tones, epichile variably longer than broad with, in some populations, a projecting pointed tip. Clinandrium (depression on top of column) much smaller than 'typical' Dune Helleborine, viscidium always absent. There is a gradation between sites that hold classic Tyne Helleborines to those with intermediate plants and those holding plants much like dune-slack populations. **VARIATION** Among dune populations, those growing in the open have a distinctive, almost 'sickly' yellow cast and often look rather wind-blasted, whereas those growing nearby in the shelter of plantations are taller, have more flowers and both the leaves and flowers are greener and 'healthier'. Dune-slack type plants growing at inland sites are also reported to be greener and more robust. **Hybrids** Hybrids with Broad-leaved Helleborine have been found in Scotland. Although confirmed by genetic analysis, this hybrid has no scientific name. At Bardykes Bing near Glasgow there is a hybrid swarm and individual plants are difficult to identify; it is debatable if there are any 'pure' Dune Helleborines present, and this may apply to the other Scottish sites.

Left: A typical dune plant, growing in Creeping Willow. 'Sickly' yellowish-green leaves are held in two distinct rows rather rigidly upwards at *c.* 45°, and sharply folded or 'keeled'. *Right:* Tyne Helleborine in birch woodland, with floppy rather than rigid, keeled leaves.

LINDISFARNE HELLEBORINE

Epipactis sancta

IDENTIFICATION

Confined to Holy Island (Lindisfarne) off the coast of Northumberland. Height 15–30cm (6.5–42cm). Identification straightforward, due to the extremely limited distribution, but very similar to Dune Helleborine, with yellowish-green leaves held in two opposite ranks and rather dull, greenish flowers which are normally self-pollinated. **Similar species** In Dune Helleborine typical dune plants have slightly larger flowers and the base of the flower stalk is washed violet (greenish in Lindisfarne Helleborine); the details of the column also differ. Lindisfarne Helleborine is even closer to 'Tyne Helleborine', which also has a greenish-yellow flower stalk and a similar column structure, but the lip of 'Tyne Helleborine' tends to be slightly longer and narrower, and does not turn under at the tip. **Flowering period** Late June and the first three weeks of July, but usually at its best early in July. The flowers go over quickly.

HABITAT

Dunes and dune slacks, especially the slightly raised and more steeply sloping zone around the perimeter of the slacks. Grows amongst Creeping Willow and various grasses or, just as frequently, on bare sand among Marram grass. May be associated with the disturbed ground around rabbit burrows.

POLLINATION & REPRODUCTION

Self-pollinated.

DEVELOPMENT & GROWTH

No information.

STATUS & CONSERVATION

Nationally Rare; listed as Endangered. For almost 50 years after their discovery in 1958, the helleborines on Holy Island were thought to be Dune Helleborines. Genetic studies show, however, that not only are Holy Island plants distinct from both typical Dune Helleborine and 'Tyne Helleborine' but also that they probably evolved independently of both of these.

150–300 spikes appear each year, singly and in small groups scattered through the dunes of the Snook at the W end of the island. Prone to drought and may not flower if it is too dry, or the buds shrivel before opening. A reduction in the number of Rabbits (which nip off the stems) may have led to an increase in the early years of the 21st century, but now it may be in slow decline: Rabbits may be important, despite their impact on flowering numbers, as they prevent scrub from invading the dunes and the bare ground created by their scrapings may help in the establishment of seedlings.

DESCRIPTION

Stem Yellowish-green, finely hairy towards tip. **Leaves** Yellowish-green, in two ranks on either side of stem, held at *c.* 45°. Lowest leaf rather short, broad and rounded, forming a cowl-shaped funnel near base of stem; 2–6 higher leaves elongated-oval, becoming narrower and more lanceolate towards spike; clearly veined and longitudinally folded but not sharply keeled. Leaf margins have tiny, very fine, regular teeth, 0.01–0.05mm wide (sometimes imperceptible). By flowering time many leaves are wind-burnt, grazed or otherwise damaged. **Spike** Fairly loose, most flowers facing to one side and held roughly horizontal or just below; the flowers droop as they age. Usually up to ten flowers, large plants may have 27 or more. **Bract** Yellowish-green, lanceolate, longer than flowers in lower part of spike but rather shorter than the uppermost leaves and becoming shorter towards tip. **Ovary** Green, ribbed, hairy and quickly inflating as flower self-pollinates. Flower stalk green or yellowish-green. **Flower** Rather small and drab but opening fairly widely. Sepals greenish, roughly triangular; petals smaller and paler. **Lip** Hypochile whitish with chocolate-brown inner rear wall to cup and sometimes pink wash to exterior. Epichile heart-shaped, tip variably deflexed; whitish, washed green in centre towards tip. **Column** Whitish; anther cap yellow with narrow brown stripe at side; pollinia ochre-yellow. Clinandria (depressions on top of column) very reduced, rostellum short, *c.* 1/2x length of anther **Subspecies** None. **Variation** None. **Hybrids** None.

GREEN-FLOWERED HELLEBORINE

Epipactis phyllanthes

IDENTIFICATION

Widespread, but local and uncommon. Height 15–50cm (5–75cm). Very variable. Some are diminutive plants with large, swollen, pear-shaped ovaries and small flowers that hang vertically downwards. Others are rather robust, with saucer-shaped flowers held facing more outwards than downwards – tall, well-developed plants being commonest in N England.

Despite the diversity, often has a fairly distinctive 'feel'. Typically relatively slender with short, apple-green leaves. Leaf shape and posture are variable, but some plants have characteristically well-spaced leaves that are very rounded and held stiffly horizontal. Flowers green with a whitish or sometimes pinkish lip, often not opening widely or, indeed, *hardly opening at all*; such plants appear to be permanently 'in bud', although the ovaries swell conspicuously. The base of the flower stalk is *greenish* and in most plants the flowers *hang vertically downwards*. Lip shape is very variable: in some populations *almost identical to the petals* (a feature shown by no other British helleborine), in others fully formed and divided as normal into a cup-shaped base and heart-shaped tip (see Variation). The flowers are generally *self-pollinated*, the pollinia crumbling in the flower onto the stigma.

Whatever the appearance of the flowers, three features confirm an identification. First, the upper stem and ovaries are either *hairless* or, rather less frequently, the upper stem may be *sparsely hairy* (use a 10x hand-lens). Second, the cup-shaped base to the lip is greenish-white, *lacking* an obvious dark brown or purple lining. Third, in difficult cases, the edge of the leaves have tiny hair-like teeth (cilia) spaced *unevenly*, in *groups* (use a 20x hand-lens).

◀ Var. *pendula (left);* Var. *degenera* (right)

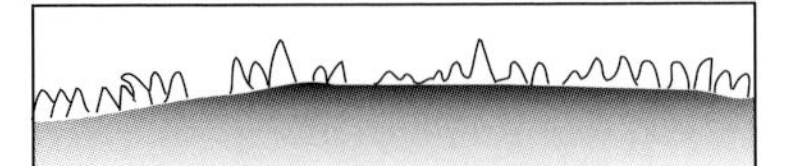
irregularly bunched teeth along leaf margin

In other helleborines they are evenly spaced. **SIMILAR SPECIES** Narrow-lipped Helleborine is also greenish overall but has a dark lining to the cup at the base of the lip and rather hairy upper stem and ovaries. Dune Helleborine also has a dark lining to the hypochile and hairy upper stem and ovaries. In addition, typical plants have a violet wash to the base of the flower stalk (yellowish-green in Green-flowered Helleborines, but also in some inland populations of Dune Helleborine – the so-called 'Tyne Helleborine'). Broad-leaved Helleborine sometimes has greenish flowers but always has a densely hairy upper stem and slightly hairy ovary. It is also cross-pollinated (see p.39). **FLOWERING PERIOD** Late June–early or even mid September but mostly from mid July–mid August, with dune populations typically earliest.

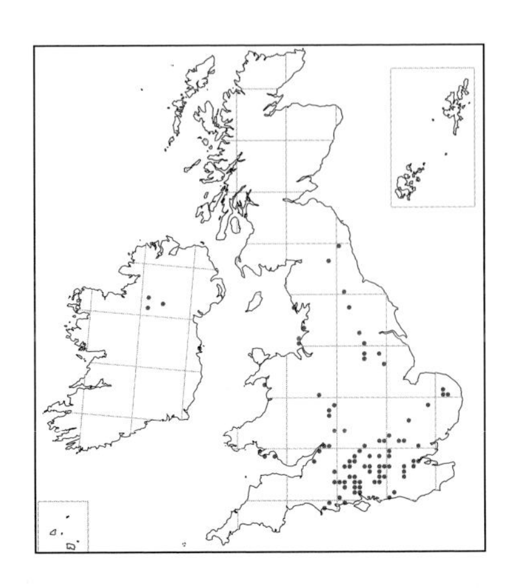

▲ Var. *vectensis*; lip turned down at tip but flowers barely opening

HABITAT

Very varied. Although it favours alkaline soils, not confined to areas of chalk or limestone and will grow on calcareous to mildly acidic sands and clays and in silty river valleys. Mostly found in light to moderate shade but can occur in densely shaded situations, although often rather small and slight in such places. Frequents a wide variety of woodlands but has a definite preference for beechwoods in S England and overall favours smaller woods, copses, belts of trees or tall hedges next to woodland, or the better-lit edges of larger woods; often found on road verges. Prefers areas where the ground cover is rather sparse or low, no taller than 15–20cm, and characteristically found growing through a carpet of Ivy.

Most of the sites for the species in S England have little or no other orchid interest. Rather than having an association with ancient woodland, many sites are relatively recent in origin, such as beech and pine plantations and shelterbelts and, in Northumberland, maturing birch and hawthorn scrub on old waste tips contaminated with zinc and lead. Another favoured habitat is thickets of willows and other trees alongside rivers and streams subject to occasional flooding, the helleborines growing on the better-drained ridges and banks. Indeed, this may be the 'natural' habitat of the species and it has even been found in Hampshire growing among reeds and willows in the tidal part of the River Itchen. Conversely, also occurs in very dry woods. At a few sites in Wales, NW England and Co. Dublin, grows in the open on sand dunes, coming up through a blanket of Creeping Willow on the drier hummocks. However, tends to look yellow and 'sickly' in this habitat, appearing rather healthier where it has spread into adjacent conifer plantations.

POLLINATION & REPRODUCTION

Always self-pollinated. In some plants this occurs when the flowers are still in bud (the flowers are thus cleistogamous); the buds may or may not open subsequently. Following pollination the whole column, together with the lip, withers rapidly but the sepals and petals can remain intact for a long time.

DEVELOPMENT & GROWTH

No information.

STATUS & CONSERVATION

Nationally Scarce. Colonies are often small, and it has a habit of coming and going at its known sites and of popping up unexpectedly in new places: the first record for Scotland came in 2015, in Moray, well outside its known range.

This species was misunderstood for a long time and widely confused with Narrow-lipped Helleborine. A better awareness of the species has led to an increase in the number of records in recent years but this is offset by the loss of known sites. Losses may be due to the grubbing-out or coniferisation of woodland, but some of its habitats are ephemeral in nature and become unsuitable as the woodland matures and becomes more shaded. In general, there has been a decrease in numbers in S England but some spread on the edges of its range.

DESCRIPTION

Underground The aerial stem grows from a rhizome that varies from nearly horizontal to nearly vertical, with numerous thick, fleshy roots, both long and short. **Stem** Usually single, occasionally 2–3 together (exceptionally as many as nine). Robust, purple-brown at base (where there are 1–3 similarly coloured sheaths) but becoming apple-green for most of its length; hairless or with sparse, short hairs. **Leaves** 3–7 (exceptionally as many as 16), apple-green, well spaced but often rather high on stem. Very variable in shape and posture but usually relatively small. In many plants broadly oval or egg-shaped but in others longer, narrower and more pointed. On all plants, upper leaves usually narrower and more bract-like, and lowest green leaf often rather small and frequently funnel-like, partially sheathing stem. Leaves may be arranged in two ranks; on some plants held flat in a very characteristic horizontal plane, on others lightly folded and positioned nearer to 45°. **Spike** Buds may be held upright, horizontally or drooping, but, once open, the flowers usually hang down near vertically; uppermost buds may fail to open. 2–25 flowers (–35),

Var. *pendula* ➤

▲ Var. *pendula*

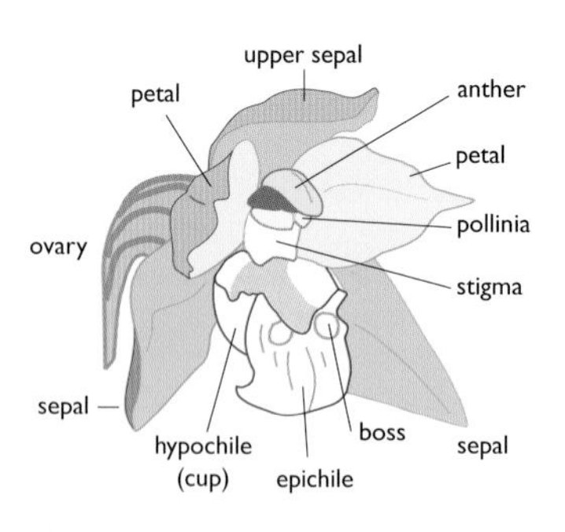

often fairly crowded, often facing in the same direction. **Bract** Green and lanceolate; lowest much longer than flowers but becoming progressively shorter – uppermost a little shorter than flowers. Bracts usually held horizontally, especially on less densely crowded spikes. **Ovary** Shiny, virtually hairless, prominently 6-ribbed, tapering into a short, curved and twisted green stalk; the ovary swells rapidly, becoming large and pear-shaped. **Flower** Var. *pendula* (see below for other flower shapes). Overall greenish. Sepals elongated-oval, variably tapering to a point, pale green with prominent green midrib on outer surface. Petals a little smaller, paler and more greenish-white. **Lip** Hypochile whitish or dull olive-white, almost translucent, with interior of cup pale greenish (sometimes lightly washed brown). Epichile heart-shaped, tip pointed, strongly turned under; two rough bosses at base frame a central groove; whitish, tinged green or sometimes pale pink in centre or towards tip, especially bosses. **Column** Whitish, anther cap dull yellowish-white with narrow brown stripes at sides. Pollinia cream or whitish, quickly crumbling and inconspicuous in anther. A small and not very sticky viscidium may be present in the bud, but withers by the time the flower opens. **Subspecies** None. **Variation** Very variable, both in flower structure and overall size and shape. The flowers range from those with a well-developed lip with a clear distinction between basal cup (hypochile) and heart-shaped tip (epichile), to those with a simple, petal-like lip. There are also parallel differences in the shape of the anther. Between the two extremes variation is almost continuous. Four named varieties, although due to the variation it can be

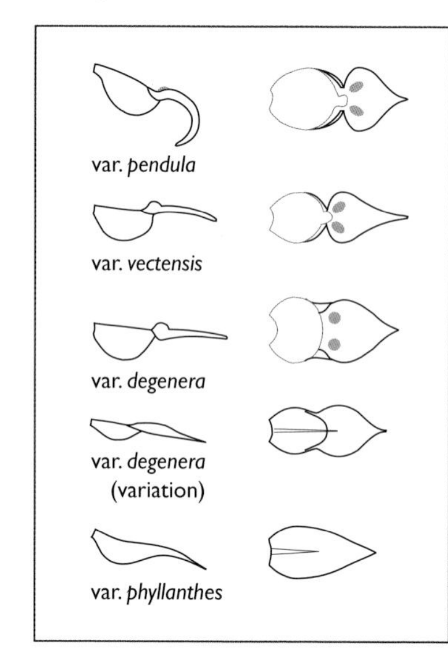

difficult to identify plants. Notably, there is no real dividing line between var. *pendula* and var. *vectensis*, and flowers with mixed characters can be found (it may be better to unite the two varieties as var. *vectensis*).

Broad geographical trends in lip shape: populations in the N and W have the most robust plants, the best developed lip and widely opening flowers (var. *pendula*); those in the south tend to be smaller, with incompletely developed flowers that open only partially. Even in the same region, however, there can be a good deal of variation, and at a few sites a mixture of lip shapes can be found in the same colony.

Var. *pendula* Lip large and fully developed. Hypochile *c.* 4mm long; epichile heart-shaped, as long as hypochile or only slightly longer, its pointed tip normally strongly turned down and under, wrinkled at base or with two bosses. Flowers open widely; cleistogamic flowers rare. The commonest variety in N Wales and N England, has been found further S; intermediates with var. *vectensis* not infrequent.

Var. vectensis As var. *pendula* but hypochile smaller, 2.5–3.5mm long, more hemispherical, embracing column closely. Epichile distinctly longer than hypochile, tip generally not reflexed – the whole lip points forwards. Flowers open to a variable extent and often cleistogamous. Mostly found in S England and the Midlands but has been recorded N to Yorks.

Var. *cambrensis* Close to var. *vectensis* but with fewer, very pale flowers and a more slender ovary; dunes in S Wales.

Var. *phyllanthes* Lip oval or lanceolate with a central midrib (i.e. not divided into hypochile and epichile), resembling a petal. Flowers rarely open widely, usually cleistogamous. Commonest in S England.

Var. *degenera* Intermediate between var. *vectensis* and var. *phyllanthes*. Hypochile reduced to a shallow depression at base of lip; there may be small bosses at sides of base of 'epichile', but lip has either no waist at all or only a rudimentary one and always lacks a central groove between the bosses. Flowers rarely open widely, usually cleistogamous. Mainly found in S England.

Size, shape and overall colour of plants also varies, especially size and shape of leaves. This variation seems to be determined by the local environment (e.g. on exposed dunes plants dwarfed and yellowish). In addition, individual plants and whole populations vary in the degree to which the flowers open and how pendulous they are, although this can change from season to season; plants in which the flower buds do not open are commonest in dry seasons, whereas in very wet years more than normal may show widely opening flowers. **Hybrids** None.

◀ Variation in lip shape

Var. *degenera* ▶

COMMON TWAYBLADE *Neottia ovata*

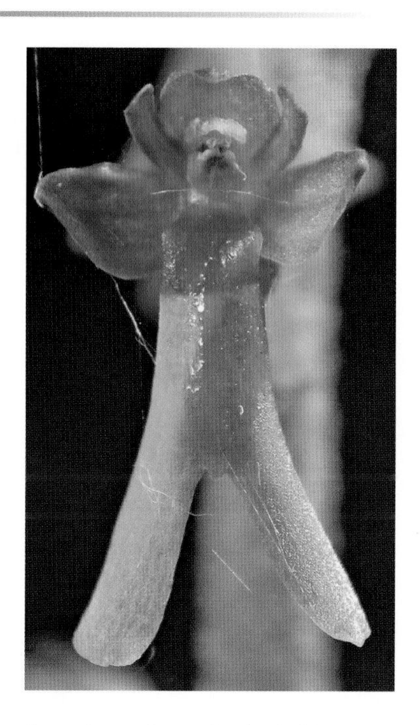

IDENTIFICATION

Widespread and locally common. Height 20–60cm (10–75cm). Identification straightforward. Green or greenish-yellow overall with two large, egg-shaped leaves placed opposite each other at the base of the stem and a tall spike of small flowers, each of which resembles a tiny green figure. Non-flowering plants, with two leaves opposite each other at the tip of the stem, are fairly frequent. **SIMILAR SPECIES** Lesser Twayblade is rather similar but tiny, with heart-shaped leaves, and flowers that are usually reddish and have sharply pointed tips to the lobes of the lip.

Man, Frog, Fen and Bog Orchids have greenish flowers but differ markedly in the structure of the lip. Man and Frog Orchids also differ in their basal rosette of leaves and Bog Orchid has tiny clasping leaves at the base of the stem. **FLOWERING PERIOD** Late April–early August, latest in the N, exceptionally even to September.

HABITAT

Possibly more varied than any other British orchid. Short chalk grassland, machair, dune slacks, limestone pavements, permanent pastures, road verges and fens, also scrub, hedgerows and moist deciduous woodland, sometimes in deep shade. Has a preference for calcareous soils but will grow in mildly acidic conditions, occasionally amongst Bracken and Heather. Can sometimes be found in relatively new habitats, such as disused railway lines, quarries and sand-pits or in plantations, even of pine. Occurs up to 670m above sea level (Perthshire).

POLLINATION & REPRODUCTION

Intensively studied by C. K. Sprengel (published in 1793) and later by Charles Darwin. Pollinated by small insects, especially ichneumon wasps, but also sawflies and beetles, attracted by the flowers' scent. Once it has landed the insect follows the nectar-filled groove up the lip and makes contact with the projecting rostellum. The pollinia are then stuck to its head by the sudden secretion or 'explosion'

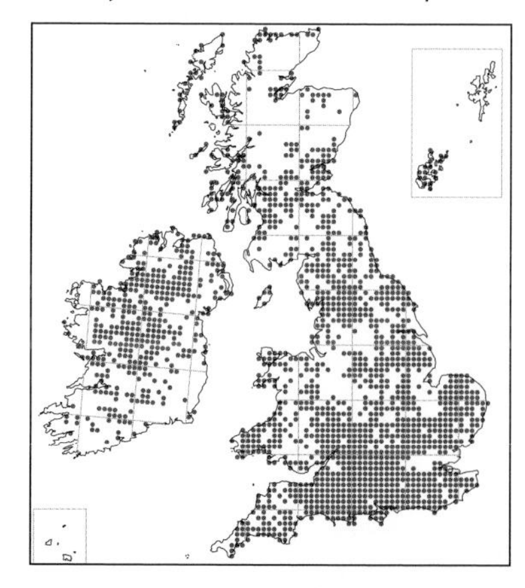

of a drop of sticky liquid that dries in just 2–3 seconds. The mechanism is extremely sensitive and the slightest touch will trigger it. The startled insect flies away with the pollinia attached, often to another plant altogether. Meanwhile, once the pollinia have been removed, the rostellum bends forwards and downwards, hindering access to the stigma and preventing self-pollination. It then slowly shrinks away upwards again to expose the stigma to visiting insects. If they are carrying pollinia these can then pollinate the flower as they try to get at the nectar at the base of the lip. The mechanism is efficient and many flowers set seed. Self-pollination may also occur occasionally; if the pollinia dry out small fragments of pollen may fall on to the stigma and effect pollination. Seed capsules contain an average of 1,240 seeds.

Vegetative propagation also occurs, with buds on the roots producing new rhizomes, and group of clones can be formed, sometimes a dense circular cluster of dozens of plants

DEVELOPMENT & GROWTH

Seeds are thought to germinate in spring, In the laboratory plants can produce leaves less than a year after germination.

Common Twayblade can be extremely long-lived. The remains of 24 old flower spikes have been counted on a single rhizome, and in a study in Sweden 20 out of 29 plants were still alive after 40 years. Mature plants have been recorded spending 1–2 years dormant underground and then reappearing.

STATUS & CONSERVATION

One of the most widespread orchids, locally common or even abundant, and usually fairly easy to find, especially in areas of chalk and limestone soils. With a very catholic choice of habitats, Common Twayblade would seem to be well-placed to survive changes to the countryside. Nevertheless, it has vanished from almost 30% of its historical range in Britain and Ireland, with a relatively large proportion of the British losses being recent.

Together with Lesser Twayblade, this species was formerly placed in the genus *Listera*, but genetic studies have indicated that this should be united with Birdsnest Orchid in the genus *Neottia*.

DESCRIPTION

UNDERGROUND The aerial stem grows from a short, thick rhizome which has large numbers of long, fairly thick roots. **Stem** Green to mid brown, thicker and whiter at base, with 2–3 membranous, scale-like basal sheaths and numerous short, white, glandular hairs. Stems usually grow singly, although not infrequently there are clusters. **Leaves** Two large, egg-shaped or elliptical sheathing leaves, held opposite each other towards base of stem; sometimes lying flat and sometimes angled at up to 45° above the horizontal; green in shade, more yellowish-green in sunny places, with 3–5 prominent veins. 1–3 tiny, triangular, bract-like leaves higher on stem. **Spike** Loose to fairly compact, with 15–30 small green flowers (–100). **Bract** Lanceolate, short; usually shorter than flower stalk and sometimes much shorter. **Ovary** Green, short, rounded, with six prominent reddish-brown ribs. Flower stalk reddish-brown, twisted and a little longer than ovary; both are variably hairy. **Flower** Sepals bluntly oval, dull green, sometimes tinged or fringed dull reddish or brown. Petals dull green, narrower and more strap-shaped; both sepals and petals curve inwards to form a very loose hood around the short, thick column. **Lip** Yellowish-green, long and strap-shaped, sharply folded down and backwards below flower and forked for about half its length into two blunt-tipped lobes; a shallow nectar-bearing groove runs down centre of lip to base of fork. **Column** Greenish. **Scent** Variously described as scented (light, musky or repellent) or odourless.

Subspecies None. **Variation Var.** ***trifoliata*** has three leaves and is not uncommon: third leaf smaller, more pointed and placed above or, less frequently, below main leaves. **Var.** ***platyglossa*** has a short lip that broadens into blunt diverging lobes, sometimes with a small tooth between them. Dunes in Donegal (where may flower till late September), formerly also S Wales. **Hybrids** None.

LESSER TWAYBLADE *Neottia cordata*

IDENTIFICATION

Occurs widely in Scotland, N England, Wales, Ireland and on Exmoor, but small and hard to find. Height 5–10cm (3–25cm; tends to be tallest in sheltered woodland). This little orchid is very distinctive, with two heart-shaped leaves set opposite each other rather high on the stem and tiny, more-or-less reddish flowers, each sitting on a large, globular ovary. On close inspection with a hand-lens the flowers resemble a tiny elfin figure. The deeply forked lip forms the legs, the two hornlike projections at its base the arms, while the sepals and petals spread star-like around the column to form a 'hat' around the 'head'. **SIMILAR SPECIES** None, but there are usually a significant number of non-flowering plants in any population, with paired leaves lying at the tip of the stem. These are very like young Bilberry plants. **FLOWERING PERIOD** Mid May–mid July, exceptionally from late April; generally peaks from late May. Once the flower has been pollinated the column quickly withers and blackens but the petals, sepals and lip sometimes persist until September.

HABITAT

Found in two, apparently distinct, habitats, but both offer the same combination of cool, humid shade and acid soils. Most frequent on wet moorland or peat bog, growing on cushions of moss, usually *Sphagnum*, under and between mature, leggy bushes of Heather, Bell Heather and Bilberry (it may be necessary to move the vegetation aside to see the orchids). The best conditions are usually found on damp, north-facing slopes. In the extreme oceanic climate of Shetland sometimes also found on short, heathy pastures. The second habitat is damp woodland, growing among a variety of mosses, sometimes in open areas and sometimes again among an understorey of Heather, Bilberry and scattered Bracken. Willow, birch and alder woods are favoured but also found in ancient 'Caledonian' pinewoods and mature pine plantations. Found up to 1,065m above sea level (Inverness-shire), and most sites are now in the hills – lowland sites have mostly been destroyed.

POLLINATION & REPRODUCTION

Pollinated by a variety of small insects, including flies and gnats, attracted by the nectar. Three pressure-sensitive hairs project from the tip of the rostellum and act as a trigger. The slightest touch by an insect causes a droplet of 'glue' to be squirted explosively onto the insect's head, and the pollinia are simultaneously released and fall onto this 'glue'. The glue dries in just a few seconds and the pollinia are carried off by the startled insect.

The pollinia are shed by the anther when the flower is still in bud and lie loose on top of the rostellum, held in position by its incurved margins. When the flower first opens the flap-like rostellum physically blocks access to the stigma and any insect visitor will trigger the mechanism the moment it touches the hairs on the rostellum. The rostellum remains in place once the pollinia have been removed but

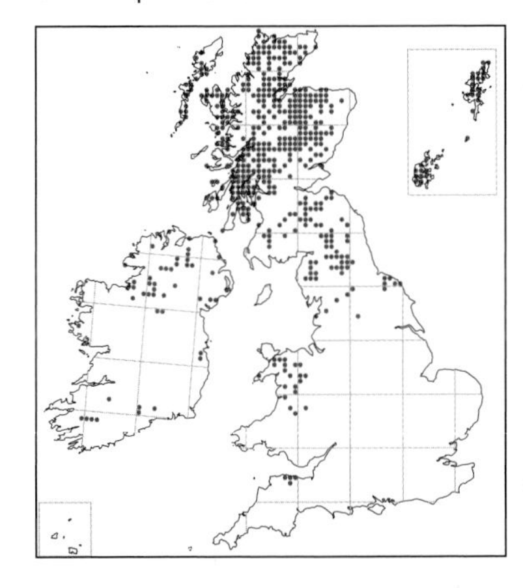

stigma or be carried there by tiny insects.

Seed-set is very efficient and capsules mature and split open within five weeks. The lowest, oldest capsules on a spike may be shedding seed before the uppermost flowers are even pollinated. Indeed, capsules swell so quickly and seed is produced so efficiently that some authors suggest self-pollination must be routine.

Lesser Twayblade also reproduces vegetatively, from buds on the roots.

DEVELOPMENT & GROWTH

Usually only one stem is produced, but buds can form at the tip of the roots and develop into additional aerial shoots that flower in their third year.

Apparently short-lived. There is little information on the interval between germination and flowering, although the first green leaf is reported to appear after 2–3 years of underground development.

is now spread flat, having released the pollinia, preventing self-pollination should the pollen-carrying insect return to the flower immediately after it has left. About 24 hours later the rostellum slowly moves upwards, allowing insects to deposit pollen from other flowers onto the stigma, which has now become very sticky. Insects are reported to work upwards from the bottom of the spike, therefore starting with the most mature flowers (i.e. those likely to have receptive stigmas), before moving on to younger flowers that still have pollinia waiting to be removed. If this is so, an insect cannot pollinate flowers on the same plant. If by any chance the pollinia are not removed by an insect, the rostellum lifts upwards anyway after a few days, giving access to pollinators.

Lesser Twayblade is self-compatible and artificial self-pollination will produce viable seed, but studies in California suggested that it is not usually self-pollinating. However, it is possible that, as in Common Twayblade, small quantities of pollen may occasionally fall from the pollinia on to the

STATUS & CONSERVATION

Locally common in Scotland, but scarcer in N England and Wales. In Ireland Lesser Twayblade is fairly widespread from Co. Sligo, Co. Cavan and Co. Down northwards but very local in the south.

An isolated population in SW England on Exmoor (Somerset/Devon), otherwise few records from S England. In Hants recorded in 1853 and 1895 near Bournemouth and in the New Forest in 1927–30 and reported again in the 1970s. Also recorded from Herts in 1980 and East Sussex *c.* 1975 and then again in 1989. Some of these records are assumed to involve plants introduced accidentally when pines or rhododendrons were planted but wind-blown seed is a possible source and in Hants there could have been relict populations on the New Forest heaths.

Has disappeared from many sites, especially in the lowlands and on the periphery of the range in N England, where the drainage and reclamation of bogs and heaths caused many losses in the 19th century. Lesser Twayblade has gone from 44.5% of its historical range in Britain and 50% in Ireland, with a rather large proportion of the losses in Britain being comparatively recent. Given that it has been better recorded in recent years because more people are actively looking for it, the actual decline has surely been substantially greater.

DESCRIPTION

UNDERGROUND The slender, creeping rhizome lies near the surface of the soil and puts out a few long, slender, hairy roots. **STEM** Green or reddish-purple, ridged towards tip, with 1–2 membranous, brownish sheaths at the base and fine glandular hairs for a short distance above the leaves. Usually grows singly but occasionally 2–3 stems grow from the same rhizome. **LEAVES** Two, opposite each other, 1/3 to 1/2 way up stem and held either horizontally or up to 45° above the horizontal. Dark, shiny green and roughly heart-shaped, with a prominent midrib that terminates in a tiny projecting point; faintly net-veined, often with undulating margins. **SPIKE** Relatively open, with 3–20 flowers. **Bract** Tiny, triangular and greenish. **OVARY** Green, spherical, with six reddish ribs, held on a reddish or greenish stalk a little longer than ovary, ribbed and twisted. **FLOWER** Very small, usually with pronounced reddish tone but can be much plainer and greener. Sepals greenish, variably washed reddish in centre and around edges, oval with blunt tips. Petals narrower, more strap-shaped, tending to be redder. Both sepals and petals are widely spread and form a star-like pattern around column. **LIP** Coppery or pale green, washed red, relatively large and triangular, deeply divided into two sharply pointed lobes. Tiny amounts of nectar are produced in a disc-shaped nectary at base of lip just below column; two very short horn-shaped lobes on either side of this nectary and a longitudinal nectar-filled groove running from it to base of fork. Lip held pointing downwards, at *c.* 90° to column.

COLUMN Short, stubby and whitish, with a large, thin, leaf-like rostellum that extends forward over base of lip, above which lie the yellow anther cap and yellow pollinia. **SCENT** A faint but unpleasant foetid odour, probably originating from the nectar. **SUBSPECIES** None. **VARIATION** **Var. *trifoliata*** has a third leaf above the usual two. Rare, but has been recorded in Scotland. **HYBRIDS** None.

BIRDSNEST ORCHID *Neottia nidus-avis*

IDENTIFICATION

Locally common in S England, scarcer and much more local in the north. Only appears above ground to flower. Height 20–40cm (15–50cm). In common with many orchids, the petals and sepals forms an open, fan-shaped hood and the lip vaguely resembles a human torso, but their honey-brown coloration is unique. Dead stems and seed capsules can remain intact for almost two years. **SIMILAR SPECIES** Yellow Birdsnest, a more-or-less similarly coloured but totally unrelated plant, is often found in the same habitats, but its flowers have 4-5 sepals and 4-5 petals, all similar in shape, and its spike curves over, only becoming erect in fruit. The various broomrapes (family Orobanchaceae) also superficially resemble Birdsnest Orchid. These chlorophyll-less parasitic plants are found in open, grassy habitats and could occur on the edge of woods or in woodland rides. Although their flowers have a 'lip', the petals are fused into a tube. **FLOWERING PERIOD** Early May–late June (late April–early July), but mostly the latter half of May. Has been recorded flowering and setting seed underground. As no systematic searches have been made, it is not known whether this is exceptional or a regular event (Australian orchids of the genus *Rhizanthella* routinely flower underground).

HABITAT

The classic habitat is the heavy shade of a mature beechwood, the orchids emerging from the leaf-litter and deep humus of a woodland floor otherwise devoid of vegetation. Also grows in mixed deciduous woodland and overgrown hazel coppice or sometimes under shady old hedges, shelter-belts or planted conifers, especially if there are still deciduous trees present. Rarely, has been recorded from grassland just outside woods. Not found in areas where the soil becomes waterlogged. Commonest on chalk and limestone soils but also grows on clays and sands that have a chalky or limestone component, such as boulder clay. Generally it is a lowland species, but has been recorded up to 250m in Cumbria.

POLLINATION & REPRODUCTION

Pollinated by insects, including flies, attracted by the nectar. The mechanism is very similar to that of Common Twayblade. The visiting insect makes contact with the projecting rostellum where there are six minute, rough, touch-sensitive points and the pollinia are stuck to its head by the sudden secretion of a drop of sticky liquid. After a while the rostellum, which has hitherto blocked access to the stigma, rises to allow visiting insects, complete with pollinia attached to their heads, to make contact with the stigma. If the mechanism is not triggered, after a few days the pollinia fragment and pollen can then fall onto the stigma below, effecting self-pollination (autogamy); pollen may also be carried to the stigma by small insects such as thrips. Ants have been noted carrying pollen from one flower to another on the same spike and this may also effect self-

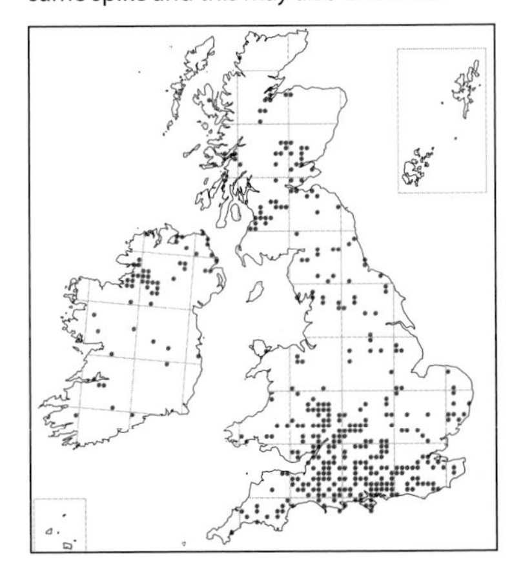

pollination (this time geitonogamy, as it is pollen from a different flower on the same plant). Occasionally, self-pollination may take place in the bud before the flowers have opened. Almost all flowers set seed.

Birdsnest Orchid is thought to be monocarpic, that is the plant dies after flowering once. But, although the rhizome dies, the numerous roots can remain alive and go on to produce new plants from buds at their tip.

DEVELOPMENT & GROWTH

A little chlorophyll is present in Birdsnest Orchid, but no effective photosynthesis takes place and it depends entirely on digesting living fungi for nutrition. In adult plants fungi are found exclusively in the roots (in the three cortical cell layers, just below the epidermis). Birdsnest Orchid is very specific about its fungal partner and only forms an association with a species of *Sebacina*, which in turn obtains its carbohydrates via a symbiotic ectomycorrhizal association with the roots of trees, particularly Beech – the tree produces carbohydrates via photosynthesis and passes these to the fungus, which in return contributes mineral nutrients to the tree. Birdsnest Orchid invades this relationship and, via the fungus, obtains nutrients from the tree. It 'cheats' in its partnership with the fungus because, unlike the tree, it does not contribute anything in return.

Seeds germinate in the spring and also require the presence of the *Sebacina* fungus. The seedling initially takes the form of a torpedo-shaped protocorm a few mm long. It then begins to develop short, fat rootlets that stick out at *c.* 90° and start to take on the appearance of a 'bird's nest'. At an early stage the bud that will produce the flower spike appears in the axil of a scale leaf at the tip of the rhizome. As the rhizome grows the number of roots progressively increases. The period from germination to flowering is probably 3-5 years.

Seed can only germinate where *Sebacina* fungus is present and this may have a localised distribution, at least partly because the fungus is entirely dependent on its host tree species – the distribution of the orchid is therefore also tied to that of the host tree. Adult Birdsnest Orchids always harbour the correct fungus and form a convenient source of infection for the seed, and germination is far more prolific in the immediate vicinity of adult plants. The clusters of Birdsnest Orchids that often grow on the exact spot that previously held a flowering plant may be the result of this, or alternatively, of the break-up of a single rhizome to produce several new plants.

Totally dependent on the activity of its fungal 'partners', not surprisingly Birdsnest Orchid does best in the warm, wet conditions beloved of fungi in general. Warm, wet springs can encourage larger numbers of plants to flower, and conversely very dry periods can result in a reduction in the above-ground population.

STATUS & CONSERVATION

Listed as Near Threatened. Birdsnest Orchid has gone from 54% of its total historical range in Britain and 45% in Ireland. Requiring heavy shade and a stable, moist environment, the species has undergone a significant decline due to the grubbing out of woodland and the conversion of deciduous woodland to conifer plantations. More subtle changes, such as widespread land drainage and the use of heavy machinery in forestry operations, could also have been detrimental.

DESCRIPTION

UNDERGROUND The aerial stem develops from a short rhizome that lies horizontally and is almost entirely surrounded by a mass of short, thick, fleshy roots that stick out at *c.* 90°. Both the English and the scientific name relate to the resemblance of this untidy mass to the nest of a Wood Pigeon or Rook. **STEM** Yellowish-brown, slightly glandular-hairy towards the tip. Grows singly, although two spikes occasionally develop from the same rhizome. **LEAVES** Green leaves absent. The lower stem is enclosed by 3–5 long, roughly oblong, scale-like, yellowish-brown, loosely sheathing leaves (the upper ones longer and blunter). **SPIKE** Cylindrical and crowded, with up to 100 flowers, but the lower flowers in the spike are usually more widely spaced and there are odd single flowers further down the stem. **BRACT** Papery, lanceolate and roughly as long as the ovary and stalk together. **OVARY** Oval, subtly 6-ribbed, glandular-hairy and held on a twisted stalk about half the length of the ovary. **FLOWER** Entirely yellowish-brown. Sepals and petals roughly oval-spatulate and form a loose fan-shaped hood over the column. **LIP** Slightly darker brown with a nectar-producing bowl-shaped depression at the base and divided towards the tip into two broad, rounded lobes that spread widely, especially on the lower flowers, to form a lyre-shape; there may also be a subtle point or tooth on either side of the lip half way towards the base. The lip is held pointing outwards and downwards at *c.* 90° to the column. **COLUMN** Pale brown, long and slender. Pollinia yellow, projecting conspicuously from beneath the anther cap. **SCENT** A pleasant but sickly, honey-like scent. **SUBSPECIES** None. **VARIATION** **Var. *pallida*** has a yellowish-white stem and flowers, and white pollinia. It is rare. **HYBRIDS** None.

GHOST ORCHID *Epipogium aphyllum*

IDENTIFICATION

Arguably the rarest wild plant in Britain and the Holy Grail for orchidophiles. Only recorded from two widely separated regions: the Chilterns and Herefordshire–Shropshire, and only seen once in the last 30 years. Only appears above ground to flower and fruit, its small size and pallid, ethereal appearance make it extremely hard to find – truly 'ghostly' qualities – and it may even flower 'underground', buried in leaf-litter. Height 5–12.5cm (–24cm). Very distinctive, pale and waxy with relatively large, pinkish flowers and no green leaves. **SIMILAR SPECIES** None. (Toothwort is ghostly white and also lacks green leaves, but otherwise very different.) **FLOWERING PERIOD** Late May–early October but mostly mid July–late September. Plants in Bucks usually flowered mid July–third week of August and those in Oxon a little later, from mid August–mid September.

HABITAT

In Herefordshire and Shropshire found in oak woodland on clay soils. Chiltern sites are in beechwoods on chalk or clay-with-flints. Usually found in heavily shaded woodland where the ground is more or less bare, often growing under a thick layer of humus or leaf-litter. Probably prefers moist but well-drained soils, thus favours the sides (rather than the bottom) of damp gullies, and has been found growing from decaying tree stumps.

POLLINATION & REPRODUCTION

Poorly-known. Bumblebees have been recorded visiting the flowers, but actual pollination has not been observed. The flowers are 'upside-down' and thus the lip does not act as a landing stage and the spur cannot contain liquid nectar (it would fall out). The flower is on a very slender stalk and the lip is reportedly 'sprung' (as in Marsh Helleborine), suggesting that bees are unlikely pollinators as they are the wrong shape and size. It is possible that other insects, such as beetles may be involved. Self-pollination in flowers in the normal pendant position is prevented by the position of the stigma, *above* the anther, because pollen cannot fall upwards. However, self-pollination may be possible while the flowers are upright when still in bud.

Whatever the mechanism, it seems that most flowers set seed, the ovaries expanding rapidly while the flower is still apparently fresh. Unfortunately, although seed ripens rapidly (*c.* 10–14 days), many flower spikes do not survive long enough for seed to mature, as they are frequently destroyed by slugs, or by rot (the stem becomes very narrow at or just below ground level, an obvious weak spot).

Vegetative reproduction may be more important than seed in producing new plants. The rhizome produces 1–2 thread-like stolons up to 50cmm long, which have bulbils every 2–3cm protected by translucent scales; it is thought that these can detach and produce new 'daughter' rhizomes, perhaps especially following soil disturbance or when washed away by flowing water.

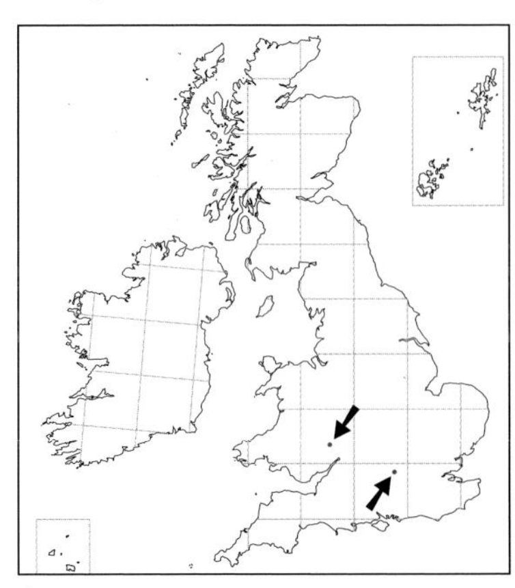

DEVELOPMENT & GROWTH

Entirely dependent on fungi for its nutrition (it is fully mycotrophic). The adult plant forms an association with basidiomycetes, probably mostly *Inocybe* sp., fungi that are ectomycorrhizal, with a symbiotic association with the roots of trees. The entire rhizome (apart from the epidermis) has a heavy, permanent presence of fungi. The thread-like stolons and bulbils are not infected with fungi, however, so if new plants develop from the bulbils, they must acquire fungus afresh.

Once the plant is mature a bud appears in the autumn and swells as the food and water reserves from the rhizome are transferred to it. If conditions are right, a flower spike may be produced the following season; it is often suggested that a wet winter followed by a warm summer may be important in stimulating flowering, but evidence is lacking.

Seed probably germinates in the autumn. No information on the length of time between germination and first flowering.

STATUS & CONSERVATION

Nationally Rare: Critically Endangered: WCA Schedule 8. Probably the hardest plant to find growing wild in Britain. Even in the right place at the right time very few flower spikes are produced.

The first British record dates from 1854, when found by the Sapey Brook on the border of Herefordshire and Worcestershire. The next records of Ghost Orchid came from Bringewood Chase near Ludlow (then in Shropshire, now administratively Herefordshire), in *c.* 1876,1878 and 1892. Also recorded from the Wye Valley near Ross-on-Wye, Herefordshire, in July 1910.

There was then a gap of 14 years before the species was seen again in Britain, this time in Oxon. In June 1924 found in woods just W of Henley-on-Thames (and seen again there in 1926, but not subsequently). In June 1931 Ghost Orchid was discovered in another complex of woods 4 miles to the SW, and it was found again there in 1933, 1953–63 and 1979. In 1953 the best-known and most productive British locality for Ghost Orchid was discovered near Marlow in Bucks, a scattered colony of 25 spikes belonging to 22 plants – the largest number ever recorded. Sightings continued near Marlow almost annually until 1987, but there have been no confirmed sightings in either Oxon or Bucks since then, despite rumours.

In September 1982, after a gap of 62 years, the Ghost Orchid reappeared in Herefordshire, when a single spike was found. Then, in a delicious irony, in September 2009, shortly after being declared 'extinct' in Britain, it was found again in the same wood.

Ghost Orchid is on the very edge of its range in England and is naturally rare. The relative lack of records in recent years may merely be part of the natural cycle of a small population or more likely reflects a genuine decline. There may be several causes: the opening of the canopy following tree falls after the great storms of 1987 and 1990 may have resulted in conditions becoming too dry for the species at some sites. The usual fungal partners, *Inocybe* sp., are scarce and may have declined due to atmospheric sulphur and other pollutants as well as fragmentation of their habitat, making the establishment of new plants less likely. Ancient woodland has been cleared or replanted with conifers (for example, Bringewood Chase in Shropshire has largely been 'coniferised', with little ancient woodland remaining). Other more direct threats include horse riders and cyclists creating paths through areas where the plants have flowered in the past, and forestry operations involving the use of heavy machinery and the dumping of materials. Some of these problems could have been avoided if owners and managers knew the location of the orchids, but such secrecy surrounded Ghost Orchid that many sightings were 'hushed up'.

Humans can be a direct threat, too, with admirers trampling the area around flower spikes. Finally, there are natural hazards, with deer and slugs taking a toll.

DESCRIPTION

UNDERGROUND Rhizome whitish, multi-branched, the branches in turn forked or tri-lobed, lobes rounded at tip and often spreading fanwise. The rhizome resembles certain corals, hence the old name, 'Spurred Coralroot'. There are no roots, rather the rhizome has a sparse covering of fine, long hairs. **STEM** Swollen at base, thick but fragile, translucent-white, washed dull rose-pink or pinkish-brown and variably streaked pinkish. Above uppermost flower stem continues as a short, narrow projection (like an aborted flower stalk). Stems arise singly but occasionally two spikes grow from the same rhizome. **LEAVES** No green leaves, merely 2–3 brown, sheathing scales at base of stem and 1–2 longer, often dark-edged, tightly clasping, scale-like leaves higher on the stem. **SPIKE** Most have 1–2 flowers, but the more robust bear 3–4; held pointing outwards or nodding downwards on short stalks. Neither stalk nor ovary twisted. **BRACT** Roughly equal in length to flower stalk and ovary together, papery and translucent, washed yellowish-straw. **OVARY** Bag-like, pale straw, veined or spotted pink or violet; flower stalk noticeably slender, curving to hold the ovary hanging downwards. **FLOWER** Spur and lip lie at top of flower, with lip facing downwards and outwards. Spur relatively large, sack-shaped, slightly curved, pale pink, filled with nectar. Sepals and petals very pale yellowish-straw with fine reddish dots or streaks; narrow and strap-shaped, the edges curled inwards, making them appear even narrower. 'Upper' sepal and two petals hang down below rear of flower, two lateral sepals project downwards and variably outwards at sides. **LIP** Whitish or very pale pink, basal hypochile short with two broad, spreading, triangular side-lobes flanking entrance to spur; epichile longer, broadly tongue- or heart-shaped, deeply concave with crimped margins and several rows of raised magenta ridges; lip strongly curved to lie parallel with spur, which it may touch. **COLUMN** Pale yellow, short, with well-developed rostellum; anther pale yellow, the two pollinia connected by a stalk at their base to two distinct viscidia. **SCENT** Nectar said to smell of fermenting bananas or vanilla but is also described as foetid. **SUBSPECIES** None. **VARIATION** None. **HYBRIDS** None.

FEN ORCHID *Liparis loeselii*

IDENTIFICATION

Very rare, confined to a few sites in Norfolk and S Wales. Height 3–18cm (–30cm). Inconspicuous and hard to find. Distinctive, the small, greenish-yellow flowers appear to be a jumble of *thin, spidery projections*. The flowers usually face upwards with the column vertical and the sepals, petals and lip all held more-or-less horizontally. **Similar species** Non-flowering plants may form the bulk of the population, and could be confused with Common Twayblade or a butterfly orchid, but are smaller and their leaves *sheathe the swollen base of the stem* to form a *pseudobulb*. **Flowering period** In Norfolk, early or mid June–early July, with a few appearing later, to late July. In S Wales, early June–late July (very exceptionally to mid September).

HABITAT

Found in marshes in the Norfolk Broads and damp dune slacks on the coast of S Wales, both offering bare ground kept damp or wet by neutral or calcium-rich ground water, low in nutrients.

In the Norfolk Broads, Fen Orchid occurs as a member of a species-rich community in areas of wet, peaty fen dominated by Common Reed or Great Fen Sedge. Grows on the mossy carpet at the base of the reeds, on the sides and tops of sedge tussocks (up to 15cm above the water level), or on the bare peat itself.

In S Wales found in dune slacks, growing with Creeping Willow, Marsh Pennywort and a variety of mosses. The slacks may be flooded in winter, sometimes for as long as five months, but the water table may fall to more than 50cm below ground level in August and September. Fen Orchid can colonise new slacks within a few years of the slack developing from bare sand and establishment from seed seems to be most successful in these young slacks. As the vegetation becomes more established, however, the orchid is less able to compete; fewer new plants appear and eventually there are no new recruits to the population. Fen Orchid will usually die out in a slack *c.* 50 years after its formation.

POLLINATION & REPRODUCTION

Probably routinely self-pollinating, the process being assisted by rain. Raindrops hit the anther cap on top of the column, which in turn knocks the pollinia towards the stigma. The upturned lip may function to deflect raindrops towards the anther. The prolific numbers of seedlings recorded suggest that seed is the major means of reproduction; in S Wales up to 128 shoots have appeared in a 0.25m^2 plot.

Vegetative reproduction also occurs but its importance relative to seed is unknown. 'Buds' (detachable propagules) are formed on the swollen stem of the pseudobulb, and these are dispersed in autumn, already carrying a fungal infection.

DEVELOPMENT & GROWTH

Seed germinates in autumn and the protocorm produces the first green leaf by the following August. The first small pseudobulb develops in the autumn of that year and the protocorm starts to

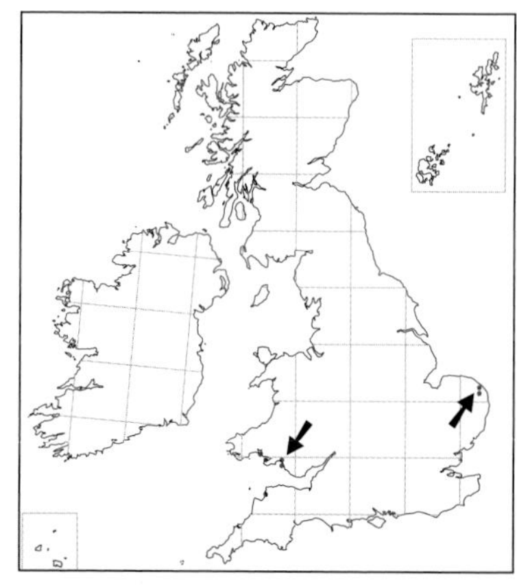

wither away. No roots are developed in the first year and all the plant's requirements for water must be met by its minute root hairs or by its fungal partner. In the summer of the second year both roots and small leaves develop and a new pseudobulb is formed. Can flower in the fourth year after germination (perhaps as early as the second year); reaches maximum size around the seventh year. There is a high mortality among immature plants and the vast majority disappear before they can flower. But, once mature, most flowering individuals reappear the next season and may flower for several years in succession; the average lifespan of mature Fen Orchids is ten years. Occasionally, plants may remain underground for a year, very rarely two, although it is hard to be sure in such cases that the leaves have not appeared and been rapidly grazed off.

Adult plants remain partially dependent on fungi, the fungal infection being concentrated in the rhizome; the pseudobulb is always fungus-free. A few short, thick, hairy roots, largely free of fungi, develop in the spring from the base of the pseudobulb.

STATUS & CONSERVATION

Nationally Rare and listed as Endangered: WCA Schedule 8. Currently five known sites holding *c.* 10,000 plants. Formerly known from 34-36 sites in E England, one in Devon and a further nine in S Wales. In E England reduced to 3–4 sites in the Norfolk Broads, where the overall population is stable: 2–3 in the Ant Valley, holding over 90% of the UK population, and a smaller colony in the Bure Valley, typically with *c.* 40 plants but up to 140 in a very good year.

Early losses in East Anglia were largely due to the drainage but although drainage and water abstraction continued to cause extinctions, most of the more recent losses have been caused by the decline of peat and turf-cutting, reed and sedge harvesting, and grazing, all traditional land uses. Even in protected areas this resulted in the development of rank vegetation, scrub and in some cases the eventual transformation of fen into wet carr woodland. Collecting also took its toll, but as with most orchids was probably a minor factor compared with habitat change.

In S Wales, Fen Orchid was first recorded in 1897 from dunes at Pembrey in Carmarthenshire and was eventually found at a further eight 'Burrows'. Still present in seven dune systems in the early 1970s but now reduced to one, Kenfig, where the population totalled over 10,000 plants until recently but has now fallen to fewer than 100. Has declined in S Wales as the habitat has become unsuitable – sites have either been buried by sand, dried out, become overgrown or have been reclaimed for heavy industry. At Kenfig, the recent huge decline is attributed to stabilisation of the dunes, which has meant that few new slacks have been formed. A lack of grazing, exacerbated by a decline in the rabbit population due to myxomatosis, has hastened the ageing process in the existing slacks. Drastic action has now been taken, scraping away vegetation and litter to produced 'young' slacks, and these have already been colonised by orchids.

The dune slack subspecies *ovata* was also found at Braunton Burrows in N Devon in 1966, but was last seen there in 1987.

In both Norfolk and South Wales the problems of Fen Orchid conservation are similar and so are the solutions. Fen Orchid is a 'weedy' species adapted to grow in dynamic, changing environments. It is able to colonise wet, calcareous habitats with plenty of bare ground and multiply quickly but is eventually crowded out as the vegetation matures. There is a high rate of turnover of individual plants, with considerable ups and downs in overall numbers. Many plants are short-lived but large numbers of seedlings can be produced, and to maintain a population conditions need to be right for seedlings to become established.

DESCRIPTION

Underground The stem grows from a pseudobulb – the swollen tip of the rhizome acts as a storage organ and is surrounded by leaf sheaths. Each summer, the rhizome continues its growth from a bud at the base of the existing pseudobulb, and this new section swells in turn as it stores nutrients. Normally, only the two youngest sections of the rhizome, representing the last two year's growth, are alive at any one time, and so there are two pseudobulbs lying side-by-side; the shiny green tip of one may be visible at base of stem. **Stem** Green, 3-angled, becoming almost winged towards tip. **Leaves** Pale green, shiny; two or very occasionally three, strap-shaped, tapering both towards base and pointed tip, prominently keeled, held erect and nearly opposite each other. Immature, non-flowering plants, with just one leaf, are common. **Spike** Loose, with 1–12 (–17) flowers. **Bract** Green, minute, lanceolate. **Ovary** Green, narrow and tubular, 6-ribbed, straight or slightly twisted at base; flower stalk 3-angled, twisted, *c.* 1/2x length of ovary. **Flower** Sepals long and narrow (*c.* 5.5mm x 1mm) the edges rolled under, making them appear even narrower; lateral sepals lie parallel below lip and are often twisted. Petals even finer (*c.* 4.5mm x 0.5mm) and usually curved, both downwards in a gentle bow and forwards in direction of lip. **Lip** Tongue-shaped, rather broader than sepals (*c.* 5mm x 2.5mm), with a wide longitudinal groove, deepest towards base, and slightly wavy or frilly margins; points upwards, parallel to column, but then bends at 90° to lie in a horizontal plane. **Column** Pale green, relatively large and prominent, slightly curved towards lip (like an erect cobra), the anther cap sitting on top and pointing forwards. Rostellum minute, pollinia two, waxy yellow, each divided into two flat plates and attached to one of two viscidia. **Subspecies** ***L. l. loeselii*** Norfolk. Up to 12 flowers. Leaves pointed and relatively narrow (at least 4x as long as broad). ***L. l. ovata*** S Wales and formerly N Devon. Averages shorter with fewer flowers (up to six, rarely as many as ten). Leaves rather more broadly elliptical or egg-shaped, blunter and more hooded at tip and held more consistently erect. There has occasionally been speculation that the two subspecies should be treated as distinct species, but genetic analysis shows virtually no difference between them, and subspecies *ovata* is sometimes treated as var. *ovata*. **Variation** None. **Hybrids** None.

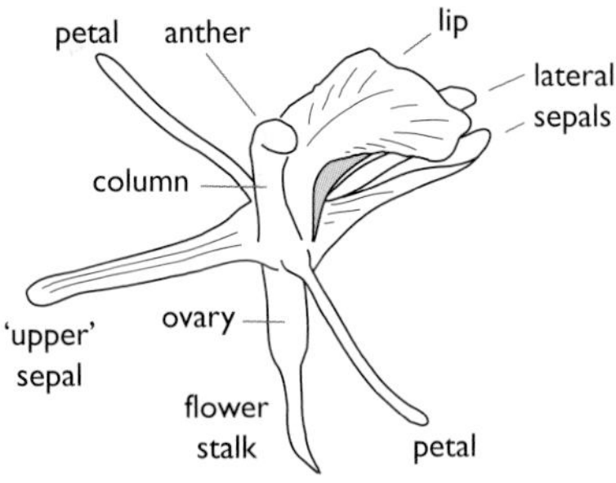

BOG ORCHID *Hammarbya paludosa*

IDENTIFICATION

Local and uncommon in NW Scotland and the New Forest; rare, with scattered sites elsewhere. Britain's smallest orchid. Notoriously hard to spot – so inconspicuous that it is easy to tread on it unawares, and the pseudobulbs, often only half-buried in the moss, can easily be dislodged; it often grows in scattered groups and 1–2 relatively obvious plants may be accompanied by several others hidden in the vegetation at one's feet. Height 4–8cm (2–15cm). The small size, tiny green flowers and habitat make this orchid distinctive – if you can find it. **Similar species** None. **Flowering period** Mid June–mid (even late) September, most reliably from early July–mid August. Tends to flower early in a hot summer and later in a cool, wet season. Numbers are erratic, with wide variations between years. Flowering at any one site is not necessarily synchronised, however, and the flowers themselves are long-lasting, thus larger colonies may have at least some plants in bloom over a lengthy period.

HABITAT

Bogs, where associated with a good cover of *Sphagnum* mosses, alongside sundews, butterworts, cottongrasses, White Beak-sedge and Cross-leaved Heath. Importantly, requires areas with a through-flow of water in the peat and avoids stagnant conditions; as well as streams and runnels, the slow flow of water may even be evident on the surface of the peat. Bog Orchid is often found in the vicinity of such moving water and also close to the shores of lakes and lochs. The ground water is usually moderately acidic but in parts of Wales the species is also recorded from areas flushed with alkaline water. It is also essential that the bog does not dry out, even in a hot summer. Bog Orchids often grow on carpets of *Sphagnum* but can also be found on bare peaty mud or in denser vegetation amidst sedges, grasses and small shrubs. Dense hummocks of *Sphagnum* are not favoured but the orchids can be found around their edges, especially if they are close to water.

Generally found in the lowlands but it has been recorded up to 500m above sea level (Caernarvonshire).

POLLINATION & REPRODUCTION

Probably pollinated by gnats and tiny flies, attracted to the nectar at the base of the lip, and the pollinia are usually removed. Seed-set is good. Perhaps more importantly in terms of reproduction, the leaves are often fringed with numerous, minute, protocorm-like buds, called bulbils. These drop off and can develop into new plants if they are infected from the soil by fungi of the right species (unlike the propagules of Fen Orchid, they do not carry fungi from the mother plant). This may account for the frequency with which Bog Orchid is found in small groups.

DEVELOPMENT & GROWTH

The stem grows from a pseudobulb that is covered by the leaves and often only half-buried in the moss. Has no roots,

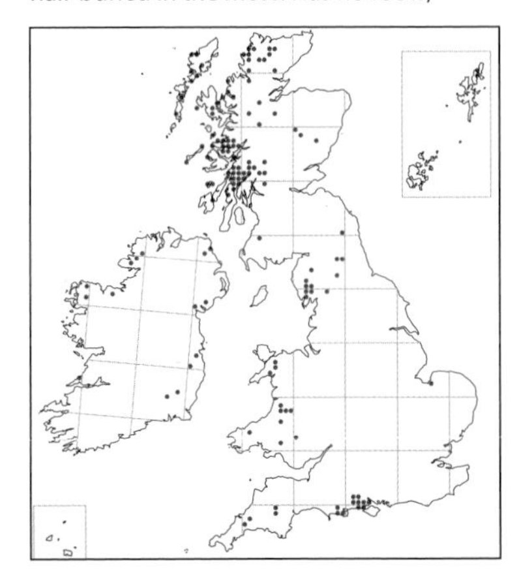

merely root hairs, and largely dependent on fungi throughout its life cycle; the rhizome and leaf bases are infected at all times. The pseudobulb, which acts primarily as a storage organ, is separated from the older parts of the rhizome by a band of hardened woody tissue. This effectively cuts off one year's growth from the next and prevents fungi from reaching the pseudobulb and its store of nutrients. An internal root is produced at the base of the new segment of the rhizome, and this grows down *through* the lignified barrier into the older segment, apparently reabsorbing nutrients and water from the decaying tissue and, bypassing the pseudobulb, carrying the fungal infection into the new segment.

Life expectancy and the period between germination and flowering are unknown.

STATUS & CONSERVATION

Threatened and declining throughout its European range due to loss of habitat, some of the largest populations are now found in W Scotland and the New Forest.

Systematic searches have produced records from many new localities in recent years but, despite this, Bog Orchid has vanished from 61% of its historical range in Britain and 66% in Ireland. Now extinct in most of England. The dramatic decline in lowland Britain started with the Enclosure Acts of the late 18th and early 19th centuries and the consequent reclamation of bogs and wet heaths. Habitat destruction has continued to the present day, and in the lowlands the remaining areas of mire and heath have also suffered from a lack of grazing, necessary to maintain the open sward that the Bog Orchid requires. Conversely, in the uplands, overgrazing may have caused suitable habitats to be degraded.

Many of the remaining colonies are small, with just a few plants. Even in its strongholds Bog Orchid is uncommon to rare: in the New Forest there are around 30 populations, with over 200 spikes appearing in good seasons at 2–3 of the largest colonies.

DESCRIPTION

UNDERGROUND The rhizome lies almost vertically in the peat or *Sphagnum*, and a swelling is formed each year at its tip. In the spring leaves develop around the base of this swelling and the flower spike grows from a bud on the top; the swollen stem and leaf sheaths combine to form

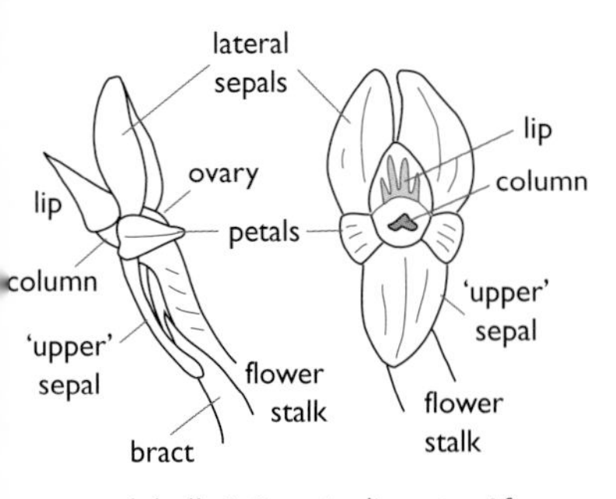

a pseudobulb 4–8mm in diameter. After it has flowered the aerial stem dies off, but the rhizome continues to grow from a bud at the base of the pseudobulb and at the end of the growing season again terminates in a swollen internode. Although in theory a whole string of pseudobulbs could be produced (as seen in many epiphytic orchids), in practice only the two most recent are alive; the older, lower, pseudobulb is buried in the moss and surrounded by the remains of the previous year's leaves. The vertical growth pattern, with the two pseudobulbs one above the other, presumably allows the orchid to adjust to the changing level of the bog's surface due to, for example, the growth of *Sphagnum*. **Stem** Yellowish-green, 3–5 angled. **Leaves** 2–3 (occasionally four), oval to oblong, prominently veined, fleshy, pale-green or yellowish-green basal leaves sheathe the pseudobulb at their base; their margins and tips are strongly curved inwards giving a hooded appearance. 1–3 minute, triangular, scale-like leaves are scattered higher along the stem **Spike** Dense at first but elongating and becoming much more open as it matures, with up to 25 flowers. **Bract** Green, narrow and pointed, about as long as ovary. **Ovary** Green, ovoid, just a little fatter than flower stalk, which is *c.* 3x its length and twisted through 360° (therefore the lip is held uppermost and the flower is termed 'hyper-resupinate'). **Flower** Greenish and tiny, *c.* 2mm wide x 4mm tall. Sepals yellowish-green, tongue-shaped; dorsal sepal points downwards, slightly longer than lateral sepals, which point upwards. Petals rather smaller and narrower, strap-shaped, green, held spreading horizontally but curving sharply back around sepals to clasp the flower. **Lip** Rather shorter than petals, triangular, curled upwards at sides. Held erect, pointing upwards and forwards between two lateral sepals, clasping column at its base; dark green with paler green longitudinal stripes. No spur. **Column** Very short, broad, green, projecting horizontally from centre of flower, the lid-like anther shrivelling to expose the pollinia shortly after the flower opens. Two pollinia, each made up of two thin plates of waxy pollen; rostellum minute, topped by a small, sticky mass. **Scent** Said to have a sweet, cucumber-like scent. **Subspecies** None. **Variation** None. **Hybrids** None.

CORALROOT ORCHID *Corallorhiza trifida*

IDENTIFICATION

Very local in N England and Scotland. Height 10–13cm (5–30cm). Small and inconspicuous, only appearing above ground to flower and set seed. Distinctive, with no green leaves (merely scale-like sheaths on the stem) and tiny whitish flowers. **Similar species** Birdsnest and Ghost Orchids also lack green leaves but Birdsnest Orchid is usually taller and always more robust, with large, honey-coloured flowers, and Ghost Orchid has a proportionally much larger flower with the lip uppermost and is only found, very, very rarely, in S England. **Flowering period** In dune slacks May–early June (often from early May), emerging earlier in drier slacks compared to wetter ones. In woodland mostly early June–late July, occasionally into August. The number of flowering spikes varies greatly from season to season.

HABITAT

Permanently damp ground with a good layer of peaty organic matter or moss, including *Sphagnum*, both in full sunlight and in shade; favours mildly acidic soils low in nutrients. Commonest in wet willow and alder carr on raised bogs and around lochs, and in damp dune slacks with a carpet of low-growing Creeping Willow; appears to have exacting requirements regarding the level of the water table and does not like prolonged flooding. Rather less frequently, found in birch and pine woods, including plantations, and in overgrown scrubby fens with a mixture of *Sphagnum*, sedges and willows. Occurs up to 365m above sea level (Aberdeenshire).

Coralroot Orchid is parasitic on a group of fungi that form mycorrhizal relationships with birches and willows (also pines in N America and presumably also in Scotland). The presence of both the correct fungal partner and one of these trees is essential and may explain its very local occurrence.

POLLINATION & REPRODUCTION

Routinely self-pollinated. The rostellum is small and degenerates quickly, and the pollinia crumble apart and fall onto the stigma below; 85–100% of flowers set seed. Small insects, including flies, wasps and beetles, visit the flowers, but the pollinia, although easily detached, do not readily stick to the insects and any cross-pollination is purely accidental; visiting insects are probably more effective in nudging fragments of the disintegrating pollinia onto the stigma below. Vegetative reproduction may also occur: the rhizome may fragment as the side-branches elongate, producing new plants.

DEVELOPMENT & GROWTH

The underground rhizome is permanently infected with fungi and, throughout its life, Coralroot Orchid is almost completely dependent on its fungal associate for nutrients. It is very fussy and only forms a relationship with the *Thelephora-Tomentella* complex of fungi (family Thelephoraceae). This group of fungi, apart from its relationship with orchids, is exclusively ectomycorrhizal, forming symbiotic relationships with the roots of trees (see Introduction). The fungi that

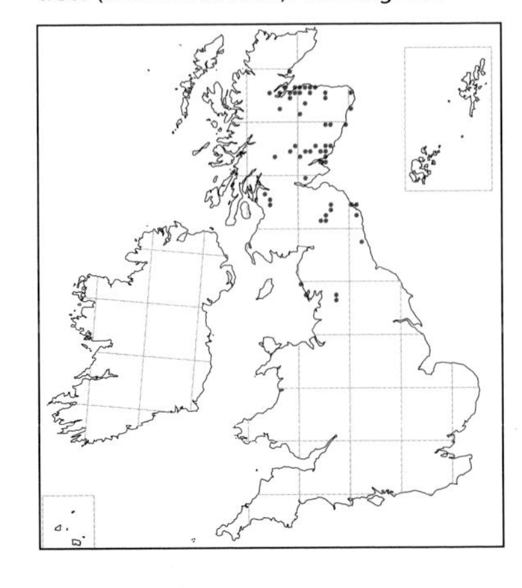

Coralroot Orchid 'partners' are species that simultaneously attach themselves to the roots of willows, birches and pines. It has been shown in the laboratory that the orchid obtains carbohydrates from the trees via their mutual fungal partner, but the orchid undoubtedly 'cheats' in its relationship with the fungus-tree partnership, receiving nutrients but giving nothing in return. It is therefore parasitic.

Coralroot Orchid does have a limited ability to photosynthesise as the stem, ovary and scale-like leaves contain chlorophyll. Even in diffuse daylight such photosynthesis can contribute in a small way to its overall nutritional budget.

Seed germinates from the spring onwards but may remain dormant until the spring of the second year, perhaps even longer. Germination will, however, only take place when the seeds have been colonised by the appropriate fungi. The seedling is initially a globular protocorm which develops scattered root hairs. It then elongates and starts to branch; each time the rhizome branches, the main rhizome bends in one direction and the side branch goes off in another, producing the 'coralloid' growth pattern. After as little as nine months, some seedlings have developed into a branched rhizome 15-25mm long with the bud for the aerial shoot already well-developed. Flower spikes may be produced 2–5 years after germination but there is a little uncertainty about the exact timing. Some related species of *Corallorhiza* in N America are monocarpic and die after flowering once.

STATUS & CONSERVATION

Nationally Scarce: Vulnerable. The true picture of its status and distribution is still emerging. Efforts have been made to find it in recent years and have turned up many new sites. Despite these new records, there has been a net loss of 46% of the total historical range and many of the losses are relatively recent. Many of its habitats seem fairly secure, however, and importantly many are not affected by changes in agricultural practices. It is possible that climate change could be responsible as many 'northern' plants are currently in retreat.

Woodland populations are usually small, scattered and hard to find but in some dune slacks there can be large numbers; at Sandscale in Cumbria at least 3,000 plants were counted in five slacks in 1991, with 1,700 in 2015, making it the largest English population.

DESCRIPTION

Underground Aerial stems grow from a creeping horizontal rhizome, a much-branched mass of cream-coloured, fleshy, coral-like knobs. There are no roots and water must be absorbed either through tufts of root hairs or via the fungal partner. Both the English and the generic name *Corallorhiza* refer to the coral-like shape of the underground rhizome. **Stem** Usually yellowish-green in woodland plants but tends to be mahogany-purple in dune populations. Frequently found in small groups and up to ten spikes may develop from one rhizome. **Leaves** 2–4 long, membranous, sheathing scales on lower half of stem that may be brown, whitish or green. **Spike** 4–9 flowers (–13) pointing outwards and slightly drooping in a lax, open spike. **Bract** Green, minute, triangular and pointed. **Ovary** Green or mahogany-purple, spindle-shaped, 6-ribbed, on an extremely short, twisted stalk. **Flower** Sepals and petals strap-shaped, petals slightly smaller; both greenish-yellow, often tinged reddish-brown around fringes and tip and thus apparently 'browned off'; interior of petals may also be blotched reddish-brown. Upper sepal and petals form a loose hood, lateral sepals curve inwards and are held forward and slightly drooping on either side of lip. **Lip** White, spotted crimson at base – the spots may occasionally coalesce into a larger blotch; tongue-shaped, shorter and broader than petals and sepals, kinked downwards towards base, with ruffled margin and tiny tooth-like side-lobes near the base; a central groove, which may produce nectar, runs between two raised longitudinal ridges; spur very short. **Column** Long, green and curved, the lid-like anther lies on top; four waxy yellow pollinia, a small rostellum and two distinct viscidia. **Scent** Slightly scented, the perfume reported to be 'musk-like'. **Subspecies** None. **Variation** Plants in dune slacks tend to be shorter than those in woodland, with reddish rather than greenish stem and ovaries, but no named varieties. **Hybrids** None.

AUTUMN LADY'S-TRESSES

Spiranthes spiralis

IDENTIFICATION

Local but sometimes abundant in S England and on the W coast north to Cumbria. Height 5–15cm (3–20cm, even 30cm). Distinctive. The small, tubular, trumpet-shaped white flowers are usually arranged in a spiral up the stem. The slender flower spikes do not grab attention, however, and can easily be missed from a walking height, while non-flowering plants are almost impossible to find; the rosettes are only 2.5–7cm in diameter, lie hidden in the grass and resemble plantains. **SIMILAR SPECIES** Irish Lady's-tresses is confined to Ireland and NW Scotland. It has long, narrow leaves on the stem and its wet grass and boggy habitat is also a good distinction. Creeping Lady's-tresses is found in pine woodland (rarely moorland or dunes), usually on acid soil, in N England and Scotland with a few populations in Norfolk. Its horizontal rhizomes form clusters of rosettes of faintly net-veined leaves with scattered flower spikes carrying exceptionally hairy bell-like flowers. **FLOWERING PERIOD** Early August–late September (early October).

HABITAT

Short, dry, nutrient-poor turf in sunny places, often near the sea and usually on calcareous soils on chalk, limestone, sand dunes, shingle banks or in the grykes of limestone pavements. Ancient earthworks are favoured, as are lawns (sometimes old tennis courts), road verges and reservoir and river embankments. Occasionally recorded from grassy places on less-acid heaths. A lack of competition from taller and more vigorous vegetation is critical, with the correct conditions provided by grazing or mowing; grazing animals may knock-over or bite off the flower spikes but ignore the leaf rosettes, and mowing also leaves the rosettes unharmed.

POLLINATION & REPRODUCTION

Pollinated by bumblebees. When the flower opens the longboat-shaped rostellum lies close to the lip at the bottom of the tubular flower, with the sticky viscidium facing downwards. In this position the rostellum blocks access to the stigma and the flower cannot be pollinated. A bumblebee lands on the lip and inserts its proboscis in search of the nectar that is produced at its base. There is just enough room for the proboscis to reach the nectaries but in the process it pushes past the rostellum, and the pollinia are attached to the proboscis by fast-drying glue on the viscidium. The bee then moves on, carrying the pollinia with it. Over the following 24 hours the column and lip move apart, creating enough space for a visiting bumblebee to insert its proboscis, with the pollinia attached, into the flower, where they rub against the stigma (which is now exposed and has become much stickier). The pollinia are brittle and break off in small pieces. Due to the 24-hour interval, older flowers are always pollinated with pollinia from a younger flower. The lowest flowers in a spike open first and, as bumblebees

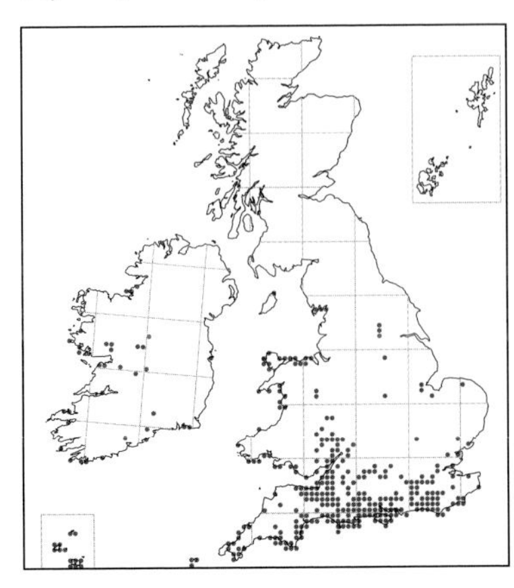

work upwards from the bottom of a spike (i.e. from older to younger flowers), they visit the older flowers first and cannot pollinate them with pollinia taken from the same spike. In this way cross-pollination is virtually guaranteed. The mechanism is efficient and seed is set by almost all flowers. The production of abundant seed is probably a factor is the lady's-tresses success in colonising new sites (although the capsules each contain an average of only 850 seeds, a relatively low figure).

Can reproduce vegetatively from lateral buds at the base of the stem, forming clusters of 2–3 plants (–12). Vegetative reproduction is, however, thought to play a minor role in the turnover of populations of Autumn Lady's-tresses.

The number of flowering spikes may fluctuate very widely from year to year, but the total number of plants present, including non-flowering rosettes or those dormant underground, is much more stable. The proportion of plants producing flower spikes is probably related to the weather. Some populations may not be able to flower for several years but will then bloom en masse when grazing or mowing ceases – in a dry summer lawns may be left uncut, allowing the orchids to flower, and until this happens it may grow unseen for many years.

DEVELOPMENT & GROWTH

Seeds germinate to produce a protocorm that is heavily infected with fungus, but by the time the roots appear the rhizome is free of fungi, although the roots are each infected in turn as they develop.

Plants may spend one or possibly more years underground with no aerial leaves and still flower the following year. This indicates that fungi play a significant part in the nutrition of the mature plant. In a study in Bedfordshire the 'half-life' averaged 6.9 years and varied from 4.6–9.2 years (see p.251).

In the laboratory green leaves are produced six months after germination, and flowering plants in five years.

STATUS & CONSERVATION

Near Threatened. Has disappeared from 55% of its historical range in Britain and 71% in Ireland. Losses are concentrated in the N and E of its former range and at inland sites in general. Sadly, has almost gone from N England, the Midlands, East Anglia and Kent.

Requiring short, nutrient-poor grassland Autumn Lady's-tresses cannot tolerate any sort of 'improvement' by the addition of fertilisers. Conversion of downland and pastures to arable or reseeding with vigorous grasses are even more directly destructive processes. Conversely, the abandonment of grazing leads to the invasion of grassland by scrub. Losses are mitigated to a small extent by gains as the species is able to colonise new sites and has appeared in large numbers on lawns in a few favoured housing estates.

DESCRIPTION

Underground The aerial stem grows from a very short rhizome that is almost concealed by 2–3 (rarely –5) thick, fleshy, tuberous roots very much like miniature parsnips in shape with a few short, transparent hairs. The roots that support the current year's leaves and flower spike eventually shrivel when their store of food is exhausted, to be replaced by new ones. **STEM** Pale green, densely covered towards the spike with fine, white glandular hairs. The flower spike grows from the centre of the *previous season's* rosette, with or without the remains of dead leaves at its base. **LEAVES** 3–7 small, narrow and bract-like, tightly sheathing the flowering stem; greenish with a narrow, whitish fringe. Just to the side of the base of the flower spike a tight rosette of up to ten leaves, dark, shiny green with a faint blue tone, oval, tapering to a point, *c.* 3cm long, with a thick keel, broadly sheathing the stem at their base. This rosette emerges August–September and overwinters, dying off late May–early June. **SPIKE** 3–21 flowers (average 9–11) arranged in a row up the stem, in most plants in a spiral, but in some the twist is so slight that the flowers form a straight line. The spiral can be either clockwise or anti-clockwise and is mostly twisted through less than 360°, although in some it may be through three full turns. **BRACT** Pale green with scattered glandular hairs towards the base and a narrow transparent-whitish fringe. Lanceolate, tapering abruptly to a fine point, a little less than 2x length of ovary, which they clasp. **OVARY** Green, 3-ribbed, stalkless, with fine glandular hairs; upright, but bent at tip – flowers held more-or-less horizontally. **FLOWER** Sepals and petals white, often washed green towards the base. Sepals oblong but tapering slightly to a blunt tip with glandular hairs on outer surface. Petals slightly shorter, rather narrower and more strap-shaped. Upper sepal, petals and lip form a long, narrow tube with the tip of the upper sepal curved upwards; lateral sepals held slightly drooped and spreading horizontally away from tube. **LIP** Pale green, whiter towards edges, with two small, globular nectaries at the base; oval or even heart-shaped with a slightly squared-off tip, sides curve upwards to form a trough or gutter; for its entire length the lip also curves downwards, the extreme tip rolled downwards and crimped. **COLUMN** Greenish, projecting horizontally into the tube made by petals and lip. **SCENT** Honey-scented. **SUBSPECIES** None. **VARIATION** None. **HYBRIDS** None.

IRISH LADY'S-TRESSES

Spiranthes romanzoffiana

IDENTIFICATION

Very local in W Scotland and N and W Ireland. Height 10–35cm. Distinctive, with *long, grass-like leaves* and a compact spike of small tubular white flowers, usually arranged into *three spirally twisted rows*. The upper stem, ovaries, bracts and sepals have numerous glandular hairs. These features, combined with its damp grassy habitat, restricted range and late flowering, are unique. **Similar species** Creeping Lady's-tresses is very occasionally found on moorland but is usually smaller, with a different flower structure and very different leaves. Autumn Lady's-tresses also flowers August–September but has a predominantly southerly distribution and is almost always found on short, dry turf. Its flowers are arranged into an obvious single row. Other white-flowered orchids, such as Small White Orchid and Heath and Common Spotted Orchids (which can have white flowers), may be found in the same habitats and in the same geographical area, but flower earlier in the season. They all lack glandular hairs on the upper stem and flowers and have a different flower structure. **Flowering period** Early July–early September, depending on weather conditions and water levels, but mostly late July–late August.

HABITAT

Open grassy areas where the soil is low in nutrients and permanently damp or wet, either because it is flushed with ground water or, more usually, because it is close to a river, stream or lake and is flooded from time to time. Often grows close to the shore of a lough and has even been recorded flowering with the leaves submerged. At most sites the water and soil are mildly acidic to neutral but sometimes recorded from more alkaline, base-rich flushes.

Suitable conditions are provided by damp meadows and rushy pastures, flushed grassy slopes, wet heathland and bogs, where it may grow amongst *Sphagnum* on the disturbed ground around old peat workings. In the Hebrides may favour the band of marginal land (known locally as 'blackland') found at the transition between lime-rich machair and acid moorland inland. And, although often growing in damp habitats, has also been found on heather moorland. 'Old lazy beds' is often given as a prime habitat for Irish Lady's-tresses (lazy beds are used to grow potatoes). Recent studies on Coll and Colonsay have, however, comparatively rarely found the species in this habitat.

Most of the sites in Britain and many in Ireland are subject to extensive grazing, which helps to keep the sward comparatively short and reduce competition from other vegetation. Disturbance by grazing animals may also stimulate dormant plants, and Irish Lady's-tresses can favour areas where cattle are fed in the winter or where the ground has been broken up by ditching or fencing. Generally found at low altitudes but recorded up to 240m in Ireland.

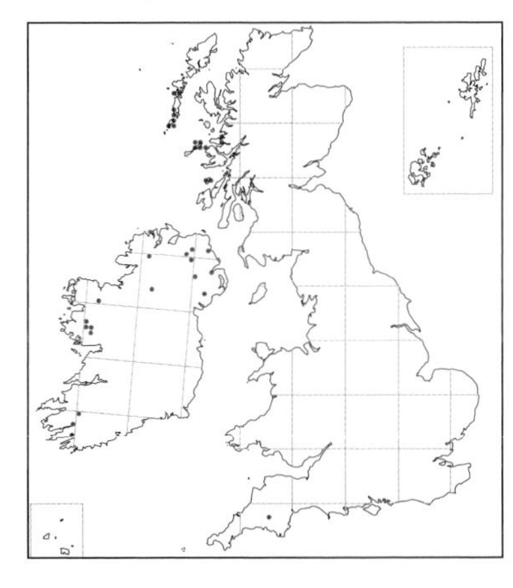

POLLINATION & REPRODUCTION

In N America pollinated by long-tongued bumblebees. The flowers contain nectar as a reward and the number of flowers setting seed is high – over 75%. The structure of the column in Irish Lady's-tresses prevents automatic self-pollination, but the flowers are self-compatible: hand pollination experiments show that self-pollination with pollinia from the same flower (autogamy) produces 64% seed set, while pollination with pollinia from other flowers on the same spike (geitonogamy) or from other plants results in 100% seed set.

By contrast, in Scotland and Ireland the reproductive biology is an enigma as plants were thought to very rarely set seed. One possibile explanation was a lack of pollinators, as bumblebees had only very rarely been recorded visiting the flowers. Concerted research efforts, however, produced more records of bumblebees visiting the flowers in Scotland, and at the largest colony in Ireland the orchids are frequently visited by medium-tongued and long-tongued bumblebees (*Bombus pascuorum* and *B. hortorum*), and occasionally by honeybees. Only bumblebees were observed to pick up pollinia, however, and presumably these would act as the primary pollinators.

Until recently ripe seed capsules had never been found in Scotland and only three times in Ireland, and it was assumed that little or no seed was produced. Careful examination of plants from Colonsay in the Inner Hebrides revealed that *c.* 40% of the flowers had their pollinia removed and *c.* 70% had pollen on their stigmas, while at the largest Irish colony pollinia were found on the proboscis/thorax of visiting bumblebees and on the stigmas of randomly checked flowers. Not surprisingly, in view of these observations, it has been shown that a small but consistent proportion of flowers on Colonsay do set seed, although many fewer than the number of flowers that are pollinated. The number of seeds per capsule is very low, but the little seed that is produced is viable and will germinate. What is surprising is that examination of more than 1,000 flowers at the Irish study site produced no ripe capsules. The reason for the very low levels of seed production is not clear, but the populations on Colonsay and in Ireland (the 'southern group', see below) have been shown to have a low level of genetic diversity, perhaps indicating a 'genetic bottleneck' during which numbers fell to a very low level. High levels of inbreeding may cause problems such as impaired stigma receptivity and low pollen viability, but examination of flowers on Colonsay showed high levels of pollen germination on the stigma, with pollen tubes growing down the style and some penetrating an ovule as normal, yet little seed develops. The northern group of populations show a high level of genetic diversity and it would seem logical that they produce seed much more consistently, but this has not (yet) been confirmed.

British and Irish populations can reproduce vegetatively through the development of an additional bud at the base of the stem (rarely two or very occasionally three). Two buds can produce two aerial stems in the next growing season, and these may eventually separate to form two plants. Extra buds are, however, only produced by a small percentage of plants each year (fewer than 5% in the Hebrides), and it seems that many of these extra buds disappear for one reason or another and relatively few develop. Nevertheless, the incidence of 'twinned' orchids in a population, presumably the product of vegetative reproduction, varies from very low to over 25% of plants. Genetic studies do not, however, indicate that groups of clones are at all common. It is possible that new plants may develop from fragments of root, perhaps broken off by cattle or sheep, but this has yet to be confirmed.

Irish Lady's-tresses is an enigma. Given its widely scattered distribution, wind-blown seed would seem the only plausible mechanism for dispersal, but seed is apparently rarely produced. Vegetative reproduction is uncommon and does not account for long-distance dispersal. Clearly, much more remains to be learned about this beautiful orchid, especially if it is to be effectively conserved.

DEVELOPMENT & GROWTH

Poorly understood. The species may become dormant underground, with up to six year's absence recorded.

Little is known about the development from seed to flowering plant, but the species possibly spends five years growing underground before the first leaves appear.

STATUS & CONSERVATION

Nationally Scarce, and a priority for conservation because Britain and Ireland hold the only European populations. Most colonies are small and scattered but some of the largest produce 50–100 spikes annually. The largest site in Scotland was recently estimated to hold 1,100 plants (flowering and vegetative combined), and 589 flowering plants were found on South Uist in 2010. Irish Lady's-tresses was found on the SW edge of Dartmoor, Devon, in 1957, but has not been seen since *c.*1993, although the site is essentially unchanged.

Irish Lady's-tresses is very hard to survey and population trends are not clear. On the one hand, it is known from an increasing number of sites, both in Ireland and Scotland, sometimes appearing well away from the known range (e.g. Kircudbrightshire in 2014). On the other hand, it tends to 'vanish' unpredictably and often rapidly from known localities. It is, however, effectively impossible to find unless in flower and the most likely explanation for its erratic appearances is that populations are relatively stable but grazing and, perhaps sometimes, the weather prevent many or most plants from flowering. Sheep, cattle or rabbits often graze-off the flower spikes and whole colonies can appear to vanish overnight if sheep are in the vicinity. As with many orchids, slugs can also graze off both the flower spikes and the leaves. Recent experiments, where sheep have been excluded from large populations, have produced a profusion of flowers, and careful long-term monitoring on Barra has indicated that the population there is comparatively stable and long-lived.

The impact of grazing on Irish Lady's-tresses is of conservation concern, but the situation is not clear-cut. Colonies, sometimes large, have been recorded on sites with a variety of grazing regimes, and indeed, it may be tolerant of heavy grazing. It is becoming clear that many orchids do best when grazing reduces competition and breaks up the sward, providing suitable sites for the establishment of seedlings, even if this means that many or most of the flowers are grazed-off before setting seed. This may apply to Irish Lady's-tresses. However, some well-established Irish sites are only lightly grazed or even ungrazed. The species is certainly vulnerable to changes in management, such as drainage, and relatively few sites are protected as SSSIs or reserves.

Irish Lady's-tresses is found in North America and on the W fringe of Europe in the British Isles. It is an extreme example of a pattern of distribution known as 'amphi-Atlantic', shared by just a few other plants. A variety of theories have been advanced to account for this distribution. It may have arrived in Britain and Ireland via long distance dispersal of wind-blown seed or perhaps as pieces of root and rhizome being carried across the Atlantic by birds. On the other hand it may be a glacial relict that once had a much wider distribution.

DESCRIPTION

Underground The aerial stem grows from a cluster of 2–6 thick, fleshy, tuberous roots, more-or-less vertical in the soil and connected at the top by a very short rhizome. A lateral bud develops at the base of the aerial stem (or on the rhizome if the plant is dormant underground) July–October and overwinters; this produces the leaves – shoots appear *c.* October, but the leaves may not expand until the spring; if the plant is to flower, the stem appears in early June. **Stem** Yellowish-green with scattered glandular hairs towards tip. **Leaves** 3–5 (–8), yellowish-green, the lower long, narrow and parallel-sided with a hooded tip, held very erect; upper short, pointed and loosely sheathing stem; leaf margins may be rolled inwards, making them even narrower and more grass-like, especially in Northern Ireland. In N America the species is wintergreen. **Spike** 5–40 flowers arranged in three rows up the stem (rarely two rows), each row variably twisted (most obvious on plants with numerous flowers). **Bract** Narrow and pointed, sheathing ovary, with glandular hairs on outer surface. Lower bracts as long as flowers, upper bracts shorter. **Ovary** Cylindrical, 3-ribbed, on a very short stalk, with a few glandular hairs. Held vertically but bends at the tip so that the flowers lie horizontally. **Flower** Sepals and petals creamy-white washed green towards base. Sepals narrowly triangular, blunt-tipped, with glandular hairs on outer surface and three greenish veins; petals narrower and more strap-shaped. Sepals and petals

form a hood that encloses the column and basal half of lip, with tips of sepals distinctly turned outwards. **Lip** Creamy-white with fine green veins, fiddle-shaped, the 'waist' nearer the tip; the sides of the larger, broader basal portion turn upwards to form a gutter, with two small nectar-producing bosses at the extreme base; smaller distal portion tongue-shaped or square-ended, frilled and toothed along edges and sharply bent downwards. **Scent** Variably strong, vanilla-like. **Subspecies** None. **Variation** Plants in Northern Ireland are said to average taller and 'leggier' than those in S Ireland, with creamy rather than white flowers, a looser spike, narrower lip and leaves often in-rolled at edges, giving them a more grass-like appearance. Plants from Scotland and Devon are intermediate, but no subspecies or varieties are usefully recognised in the British Isles. However, genetic studies have shown that there is a split between a northern group of populations, from Coll, and from Barra and Vatersay in the Outer Hebrides and a southern group from Colonsay and Ireland (no plants from mainland Scotland were examined). There has been little recent contact or gene flow between these two groups, but this division is not reflected in any known differences in appearance or ecology. **Hybrids** None.

CREEPING LADY'S-TRESSES *Goodyera repens*

IDENTIFICATION

Locally common in Scotland, with populations of uncertain status in N England and Norfolk. Height 7–20cm (–35cm). The spikes of small, densely hairy, white flowers are distinctive. The only evergreen British orchid, it can be found and identified year-round. Indeed, often easier to locate in winter when other vegetation has died down. Forms small clusters of rosettes composed of dark-green leaves rather like Garden Privet in size, shape and colour. Notably, the leaf veins form a *faint net* over the surface (almost all other British orchids have *parallel* veins). **Similar species** Autumn and Irish Lady's-tresses are in a different genus, *Spiranthes,* but are nevertheless rather similar, having small white flowers with glandular hairs, although neither are as densely hairy as Creeping Lady's-tresses. The flower structure is similar but the two *Spiranthes* have the tip of the lip broadly frilled or crimped, rather than being a simple wedge. Autumn Lady's-tresses is unlikely to be found in pinewoods. Its flowers are arranged into a neat row that is often (but not always) spirally twisted. Irish Lady's-tresses grows from a rosette of long, strap-shaped leaves and its flowers are arranged into a more definite pattern. **Flowering period** Late June–late August; often at its best in mid July.

HABITAT

Mature pinewoods with a damp, well-shaded forest floor and deep humus formed by the accumulation of dead pine needles. Found in areas with an open understorey of grasses and small shrubs, such as Heather and Bilberry, or where there is merely a carpet of moss over the needles (sometimes even bog mosses, *Sphagnum*). Its classic habitat is ancient Caledonian woodland of Scots Pine mixed with birches, but it has also spread into well-grown pine plantations with a shaded forest floor. A balance of light conditions is required; too little and the orchid will not flower, too much and a dense, shrub layer develops. Rarely, found on damp dunes or on moorland among Heather and Bell Heather, sometimes far from woodland (perhaps most often in coastal areas). Unlike most British orchids, grows on acid as well as neutral soils. Mostly lowland, but recorded up to 335m above sea level (Banffshire).

POLLINATION & REPRODUCTION

Thought to be pollinated by bees, which are attracted to nectar at the base of the lip. Bumblebees of the genus *Bombus* have been recorded carrying pollinia, but smaller bees of the genus *Lasioglossum* may be more important. The pollination mechanism is probably similar to that of Autumn Lady's-tresses. The flower tube is initially only wide enough for an insect's proboscis to enter; this trips the mechanism and removes the pollinia. The lip then moves slowly downwards, allowing access to the stigma, which becomes sticky and receptive. Fragments of pollinia, picked up from other flowers and attached to a visiting insect's

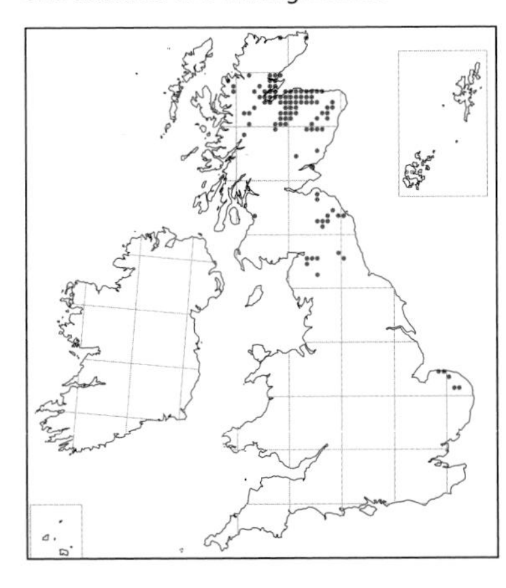

proboscis, can then effect pollination. Seed-set is good, with 77% of flowers setting seed in one Scottish study.

Vegetative reproduction may be more important than reproduction from seed. Buds are produced on the tip of the rhizome in the autumn, and these grow into slender runners that grow horizontally through the cushion of moss. At first these have merely a few short, sheathing scales, but eventually green leaves are produced from buds at or near their tips. Once it has appeared above ground, the new rosette will live for 2–8 years before flowering and dying (in Norfolk a period of at least six years has been recorded between the first appearance of rosettes and the production of flowers). Each runner can produce a separate plant; after flowering the central 'mother' plant dies off, leaving the surrounding rooted runners as separate entities. In this way large patches can form.

DEVELOPMENT & GROWTH

Both rhizome and roots have a heavy presence of fungi and the species is probably dependent on its fungal partner to a significant extent throughout its life. Unlike most other orchids, however, this species shows no tendency to disappear below ground and assume an existence totally dependent on fungi.

Seeds probably germinate in spring. The period between germination and the first appearance above ground is not known.

STATUS & CONSERVATION

Common to locally abundant in N and N-central Scotland, especially the Strathspey and Cairngorm regions; scarcer in W Scotland and N England. Colonial, it may occur in large numbers; a Norfolk colony had 620 flower spikes in *c.* 50m x 20m.

Although still locally common, there has been a significant reduction in the range, and it has now gone from 44% of its historical distribution. With its requirement for a minimum level of shade, vulnerable to woodland management. If a wood is felled and replanted, colonies may survive, but it takes many years for the number of flowering plants to recover and they must be replanted with pines; many sites have been lost when restocked with alien conifers such as spruce and fir. Heavy thinning can also have a severe effect, although perhaps not permanently so.

The status of the species in Norfolk has been the subject of debate since its discovery in 1885. Scots Pine is not native to Norfolk, but there are conifer plantations dating from at least the 1830s. It has been widely assumed that Creeping Lady's-tresses was accidentally introduced with pine seedlings brought from Scotland, but there is no evidence to support this. An alternative theory is that it was present on heathland in Norfolk before the advent of pine plantations and merely moved in when they became suitable; it was found at Sheringham in 1909, concealed among Heather on open heathland, and found on heathland at Beeston Regis in 1900. Alternatively, Norfolk populations may originate from wind-blown seed, either from Scotland or the Continent. The timing of records is interesting, as Creeping Lady's-tresses were recorded for the first time from some plantations many years after they were established; pines were first planted at Wells and Holkham in the 1850s, largely with seedlings produced in a nursery at Holkham itself, but lady's-tresses was not found there until 1952. Whatever their origin, the populations in Norfolk are notable from a scientific point of view, yet doubt over their status has compromised their conservation. There are old records for SE Yorkshire and E Suffolk (1932–35, Stuston Common, a site with no pine trees).

DESCRIPTION

Underground The aerial stem grows from a creeping rhizome, often only shallowly buried, with a few short, thick, fleshy, hairy roots. **Stem** Pale green, ridged, with dense glandular hairs towards tip. **Leaves** Dark green with a network of faint paler veins (reticulations); 3-9 form a rosette, more-or-less flat to the ground, there may also be 1–2 at base of flower spike; oval, tapering to a point, leaf-stalk short but broad. Several very small, long, narrow sheathing leaves grow further up stem. Sterile, non-flowering rosettes also produced. **Spike** 5–25 flowers arranged spirally around stem (spiral seldom obvious - they mostly face in the same direction). **Bract** Green, with scattered glandular hairs towards base; lanceolate, slightly longer than ovary, which they clasp. **Ovary** Green, narrowly pear-shaped, tapering towards base, 3-ribbed, with glandular hairs. Upright, but bends at tip – the flowers lie either horizontally or slightly drooped – and either stalkless and slightly twisted or with a short stalk, in which case stalk slightly twisted. **Flower** Small, creamy-white and densely hairy. Sepals oval, concave, upper sepal a little narrower than lateral sepals; they may have a faintly greener midrib and have many conspicuous long, glandular hairs on the outer surface. Petals similar but slightly smaller and more spatulate; upper sepal and petals form a tight hood, lateral sepals held drooped and slightly spreading. **Lip** White, shorter than sepals, oval, pointed; basal hypochile forms a deep, rounded, bag-shaped pouch that contains nectars and is sometimes tinged pink; epichile a narrow, blunt-tipped triangle, folded into a shallow groove and bent downwards towards tip. **Column** Creamy, rostellum formed into short, curved horns that enclose the single, roughly circular viscidium. Anther cap ochre with a reddish-brown margin, pollinia yellowish. **Scent** Sweet. **Subspecies** None. **Variation** None. **Hybrids** None.

MUSK ORCHID *Herminium monorchis*

IDENTIFICATION

Scarce and very local in S England. Height 3.5–15cm (2–20cm, rarely to 30cm). Small and easily overlooked, but its crowded spike of tiny greenish-yellow flowers is distinctive. Unless examined closely, the lip is hardly different from the petals and sepals and the flower appears to be made up of six nearly identical narrow 'petals', forming a little bell. **SIMILAR SPECIES** Bog Orchid is also very small and greenish-yellow but is strictly confined to acid, boggy ground, and the structure of its flower is different. **FLOWERING PERIOD** Early June–early July (occasionally early August).

HABITAT

Found exclusively on short, well-drained grassland on chalk or limestone soils. Its small stature means that it cannot compete if the vegetation is tall, so thin or compacted soils that restrict plant growth are favoured. It particularly likes the narrow terraces formed on steep downland slopes by soil creep, as well as ancient earthworks, abandoned quarries, chalk and lime pits, and spoil heaps.

POLLINATION & REPRODUCTION

Pollinated by a variety of tiny insects such as flies, parasitic wasps, gnats and beetles, typically just 1–1.5mm long. As it feeds on nectar at the rear of the lip, the insect ruptures the skin of the relatively large viscidia, and these stick the pollinia to its legs. Once the insect leaves the flower, the pollinia rotate forward to be in the correct position to make contact with the stigma of the next flower to be visited; this can be on the same plant. There are conflicting reports on the possibility of self-pollination. A study in Sweden found that it did not take place, but it has also been stated that the flowers can be self-pollinated, the anther withering and the pollinia dropping onto the stigma immediately below. Whatever the mechanism, 70–95% of flowers set seed.

Vegetative reproduction may be the major means of recruitment to the population; 2–5 'daughter' tubers are produced at the ends of slender rhizomes up to 20cm from the parent tuber. In this way extensive clones can develop.

DEVELOPMENT & GROWTH

The tip of the tuber and usually also the roots are infected with fungi. Two or more new tubers are formed each growing season, the larger of which provides for the following year's leaves and flower spike, the smaller are 'daughter' tubers. It is sometimes described as 'migratory' because the replacement tuber grows at the end of a short rhizome and therefore the aerial shoot appears in a slightly different place each season. After flowering once, plants may appear again merely as vegetative rosettes or even remain dormant underground for 1–2 years before flowering again.

There can be large variations from year to year in the number of flowering spikes, with the temperature and amount of rainfall over the previous summer being the determining factors. Despite its

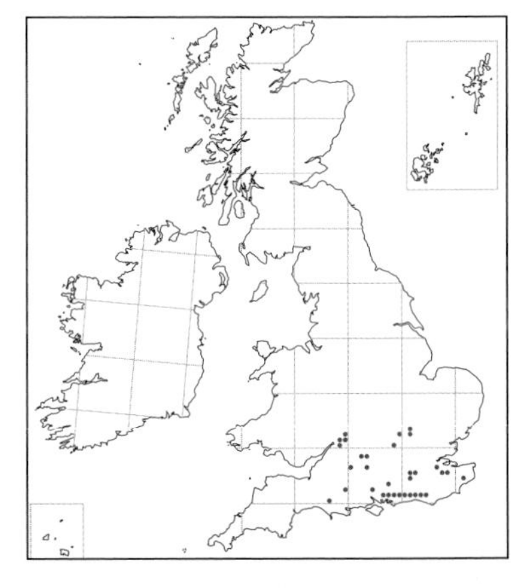

predilection for dry chalk grassland, it is vulnerable to summer drought, which can lead to the early withering of the leaves and thus a reduction in the amount of food resources stored in the tubers for the following season. A hot dry summer can thus cause a big fall in the number of flowers the following year, the plants appearing merely as non-flowering rosettes. Populations can recover rapidly following a good growing season, but prolonged droughts can be devastating.

Reported to flower after a period of immaturity lasting several years, but in cultivation plants have flowered within two years of seed being sown. Individuals can be long-lived, with an age of 27 years being recorded.

STATUS & CONSERVATION

Nationally Scarce and listed as Vulnerable. Confined to the N and S Downs in Kent, Sussex and Surrey, also Hants, Dorset (just two sites), Wilts and Berks, the Cotswolds in Glos and the Chilterns in Bucks, Herts and Beds. Very local and is absent from large areas of apparently suitable habitat. At a few favoured sites occurs in large numbers and can form dense and obvious stands. Otherwise hard to spot, especially on anything but the shortest turf or if the spikes are few and scattered. It can be easier to tread on than to see, so it is good practice to keep to whatever paths are available.

Musk Orchid has been lost from 69% of its total historical range and was recorded from just 32 10km squares in the *New Atlas* period of 1987–99. The ploughing of chalk grasslands from the late 18th century onwards caused the first wave of losses, especially in E Anglia, and many sites had gone by 1930. Subsequently, the 'usual suspects' of agricultural intensification, the scrubbing-over of grassland (especially following the outbreak of myxomatosis) and overgrazing have taken their toll. The species can, however, at least occasionally, colonise new areas.

DESCRIPTION

Underground Aerial stems grow from a single spherical tuber that starts to wither away by flowering time; there are a few short, thin roots. **Stem** Yellowish-green to dark green, distinctly ridged towards tip. **Leaves** Two mid green basal leaves (rarely 3–4), strongly keeled, oblong to oval-oblong, and 1–3 small, lanceolate, bract-like leaves further up stem. Leaves emerge from early May onwards and persist until mid September, but in a dry season they may not appear until early June and wither early. **Spike** Often 1-sided, with 20–30 flowers (over 70 on the very largest); appears densely packed due in part to the relatively large ovary. **Bract** Green, lanceolate, roughly length of ovary. **Ovary** Relatively large and inflated, even when flowers fresh, greenish-yellow, prominently ribbed and slightly twisted. Upright, but narrows at top into stalk-like base to the flower that is bent through more than 90°, holding the flower pendant. **Flower** Greenish-yellow and very small, *c.* 2.5mm wide x 3mm front–back. Does not open widely – the sepals, petal and lip all point forward and are more-or-less parallel, giving a spiky, tubular or bell-like shape. Sepals oval, lateral sepals slightly smaller than upper sepal. Petals slightly paler and, rather unusually, longer than sepals; spear-shaped, with variable small side-lobes towards base and narrower tip. **Lip** Strap-shaped with a long, narrow central lobe and rather shorter side-lobes; narrows at rear into a short, blunt chamber (a rudimentary spur) that secretes nectar. **Column** Two oval pollinia, each attached by a very short, elastic caudicle (stalk) to a saddle-shaped viscidium almost the same size as the pollinia. The viscidia each have a delicate skin but are not enclosed in a bursicle (pouch). **Scent** Sweet and honey-like. **Subspecies** None. **Variation** None. **Hybrids** None.

GREATER BUTTERFLY ORCHID
Platanthera chlorantha

IDENTIFICATION

Found locally throughout Britain and Ireland. Height 20–40cm (–65cm). The two butterfly orchids are distinctive, with a pair of oval, shiny green leaves at the base of the stem and an open spike of beautiful 'waxy' white or greenish-white flowers. The lip is long, narrow and undivided, and the extremely long slender spur projects backwards from the rear of the flower across the width of the flower spike. **SIMILAR SPECIES** Lesser Butterfly Orchid is distinguished by the shape and position of its pollinia, which *lie close together and are parallel for their entire length*. In Greater Butterfly Orchid the pollinia are *well-separated at the base but lean inwards so that their tips almost touch*. Greater Butterfly Orchid is on average taller and sturdier, with a broader flower spike (the ovaries are longer, holding the flowers further away from the stem). It has greener flowers with a larger and more obvious mouth to the spur, and the spur itself is usually slightly expanded at the tip. The two butterfly orchids, although readily separated by the shape of the pollinia, are genetically almost indistinguishable, suggesting that they have only very recently separated into two species. **FLOWERING PERIOD** Late May–late July, earliest in the S, where it is sometimes in flower from mid or even early May. Usually it is at its best in early June in much of England but it may not be in full flower in parts of Scotland until July. (Typically 1–2 weeks earlier than Lesser Butterfly Orchid.)

HABITAT

Rather variable but almost always on calcareous soils: chalk, limestone and base-rich clays. Deciduous woodland (where strongly associated with ancient woodland), with a preference for hazel coppice. It grows in light, dappled shade and is usually found in the more open areas around the edge of a wood and in clearings and along rides. It is tolerant of both dry and wet conditions but in woodland on chalk may have some preference for the heavier and wetter soils found at the foot of slopes. Has occasionally been found in rides through conifer plantations but such populations

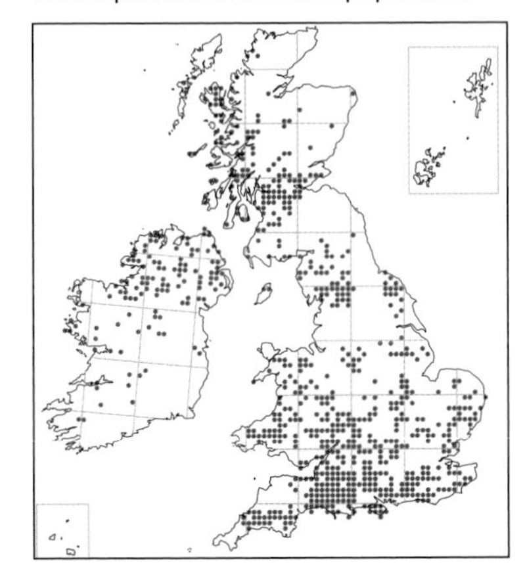

may be relicts of deciduous woodland previously on the site.

In N and W Britain and Ireland most frequently found in old pastures and hay meadows, again usually on calcareous soils; on rare occasions found on mildly acidic soils on moorland, wet heath and pastures but less tolerant of acid conditions than Lesser Butterfly Orchid. Probably frequent in pastures and meadows in S England prior to their 'improvement' but away from woodland sites now only likely to be found on chalk grassland, often in long grass with variable amounts of scrub. Sometimes also found in calcium-rich dune slacks and on railway embankments. Occurs up to 460m above sea level (Northumberland).

POLLINATION & REPRODUCTION

Night-flying moths are the primary pollinators, principally the larger, longer-tongued members of the Noctuid family (occasionally also hawkmoths). The sticky viscidia at the base of the pollinia face inwards on either side of, and just above, the entrance to the spur. Being relatively widely spaced, the pollinia usually become attached to the large compound eyes of a visiting moth. After *c.* 2 minutes, the pollinia's stalks pivot so that the pollinia face forward and are correctly positioned to make contact with the stigmatic zone on the next flower visited. The mechanism is effective and seed is set by *c.* 45–90% of flowers. (See also Lesser Butterfly Orchid.)

Vegetative reproduction can occur via the formation of additional tubers but is of little importance.

DEVELOPMENT & GROWTH

The process of development from seed to flowering plant is very similar to Lesser Butterfly Orchid. The adult plant is reported to be entirely independent of fungi, but this seems highly unlikely given that it can become dormant in heavy shade for long periods. In many populations a large proportion of plants are non-flowering, and when heavily shaded in overgrown coppice or very dark woods plants can remain in a vegetative state for decades. They 'reappear' and flower again following coppicing, tree-falls or other changes that let more light in. In woods with a regular coppice-cycle, flowering is most prolific 2–3 years after coppicing.

STATUS & CONSERVATION

Near Threatened. Has been lost from 46% of the historical range in England and 53.5% in Ireland. The destruction and the replanting of woodlands with conifers, as well as agricultural improvements to pastures and hay meadows, have been responsible for the decline.

DESCRIPTION

Underground The aerial stem grows from a pair of spindle-shaped tubers that taper to a long, narrow point; there are also a few slender roots that spread into the surface layers of the soil. **Stem** Pale green, more-or-less triangular and ribbed towards tip, with 1–3 brownish sheaths at extreme base. **Leaves** Two, oval to elliptical, keeled, pale green (often slightly bluish) and rather shiny. Held variably erect, they lie opposite each other at base of stem, one just above the other;1–6 small, lanceolate, bract-like leaves higher on stem. **Spike** Variable. 10–30 (–40) flowers form a loose, open spike in woodland plants; in full sun tends to be more compact. **Bract** Green, narrow, pointed, as long as ovary. **Ovary** Pale green, long, narrow, clearly 6-ribbed and twisted, and curved to hold flowers pointing outwards and a little downwards.

Flower Sepals white, washed greenish towards tip; lateral sepals oval-triangular, asymmetrical and even sickle-shaped; upper sepal rather shorter, broader and more triangular or heart-shaped. Petals shorter, narrower and more strap-shaped than sepals; upper sepal and petals form a loose hood over column, lateral sepals spread horizontally and are often a little twisted or wavy-edged.

Lip Creamy but becoming greener towards tip, narrow and strap-shaped, 10–16mm long, projecting forwards and downwards. Spur long and slender, 19–35mm long x 1mm wide, broader towards tip and often strongly curved; washed green.

Column Pollinia yellow, relatively large, 3–4mm tall including the long caudicle (stalk). Viscidia at base of pollinia lie *c.* 4mm apart on either side of foot of column; pollinia lean inwards towards each other.

Scent Emits a heavy scent, especially at night; it is said that people with a sensitive nose can smell them from several hundred metres on a still summer's evening.

Subspecies None. **Variation** Plants growing in the open tend to be shorter and more compact and to have shorter leaves than those in shaded woodland localities. It seems likely that these differences are produced by the local environment rather than by any genetic difference. Plants may occasionally have just one leaf or sometimes 3–4. **Hybrids** See Lesser Butterfly Orchid.

LESSER BUTTERFLY ORCHID

Platanthera bifolia

IDENTIFICATION

Found throughout Britain and Ireland although rare or absent in SE England, the Midlands and N England. Height 15–30cm, (–45cm). Butterfly orchids have two oval, shiny leaves at the base of the stem (often hidden in the grass), and beautiful white flowers with a long strap-shaped lip and extremely long and slender spur that projects prominently behind the flower. **Similar species** Lesser Butterfly is the more delicate of the two butterfly orchids, averaging smaller and daintier (*c.* 2/3 the size of Greater Butterfly Orchid in most dimensions), with fewer, smaller flowers in a narrower spike; the mouth of the spur is smaller and the spur tends to be straighter. The only diagnostic distinction is, however, the size and shape of the pollinia. In Lesser Butterfly they are *placed close together and parallel*, while in Greater Butterfly their bases are well separated and the pollinia *lean in towards each-other.* The two species are extremely close genetically, suggesting that they have only very recently separated. **Flowering period** Late May–June in woodland populations, June–July for heathland forms. Averages latest in N Scotland.

HABITAT

The 'heathland' form of Lesser Butterfly Orchid is by far the commoner and grows on heathland in S and E England and on moorland and damp pastures in the N and W. Usually grows in damper areas, often around the margins of valley bogs or growing on drier tussocks and other raised areas amidst sodden, boggy ground. It grows on neutral or mildly acidic soils – on moorland and around bogs favours the areas where springs and flushes buffer the general acidity. The 'woodland' form is much scarcer and largely restricted to S England, in open deciduous woodland and scrub, often beechwoods, on calcareous soils, especially 'clay-with-flints'. It can also occur on open chalk downland and among Bracken on the less acid areas of dry grassy heaths. Recorded up to 365m above sea level (Inverness-shire).

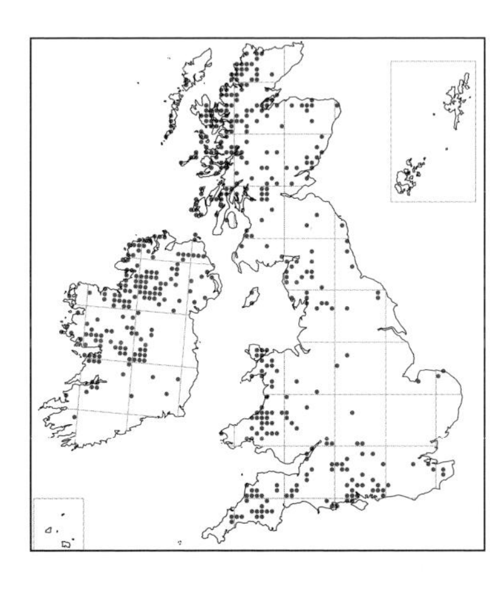

STATUS & CONSERVATION

Listed as Vulnerable. Has declined greatly, with a loss of 64% of the historical range in Britain and 48.5% in Ireland, and has vanished from much of S and E England. It is also seriously reduced in many other areas. For example, in Norfolk it was 'locally common' on wet heaths in 1914 but only 'fairly frequent on bog and wet heath, always on acid soils' in 1968, when 12 localities were listed, and by 2015 Lesser Butterfly Orchid was present at only 1–2 sites with just a handful of plants. It has now gone altogether from some classic heathland area in S England such as Woolmer Forest in Hants and Ashdown Forest in Sussex.

Some losses are due to the outright destruction of heathland, with both agriculture and urban development being responsible. Even where heathland remains, the losses continue, probably due to the lack of grazing and resulting transformation of heathland into scrub and woodland. Notably, Lesser Butterfly Orchid is still common in the New Forest, which continues to support large numbers of grazing animals. Away from heathland sites the 'improvement' of pastures and hay meadows and the clearance or 'coniferisation' of woodland are responsible for the decline. Paradoxically, overgrazing is a problem in the uplands.

POLLINATION & REPRODUCTION

Pollinated by night-flying moths, especially hawkmoths such as Elephant, Small Elephant and Pine Hawkmoths, attracted by the copious nectar in the spur (nectar can be seen filling the tip of the spur, which is translucent); the nectar can only be reached by an insect with a suitably long proboscis. Moths are drawn by the flowers' scent, which is particularly pungent around dusk, and white coloration (they almost 'glow in the dark'). Hawkmoths hover in front of the flower to feed, resting their forelegs on the lateral sepals, whereas other moths land on the flower itself. The two pollinia, which are relatively small, extremely short-stalked and placed parallel to each other at the mouth of the spur, have small sticky pads, the viscidia, at their base. As the moth inserts its proboscis into the spur, the viscidia glue the pollinia to it. The moth continues on its way, visiting other flowers, and after a short while the pollinia rotate forwards and in this new position will make contact with the stigma in the next flower visited. The mechanism is fairly effective and seed-set is moderate to good.

DEVELOPMENT & GROWTH

No information on the period between germination and flowering in the wild; in cultivation plants may flower in 3–4 years.

DESCRIPTION

Underground Grows from a pair of underground tubers. **Stem** Pale green, more-or-less triangular and ribbed towards tip, with 2–3 whitish or brownish sheaths at extreme base. **Leaves** Two pale green, slightly shiny leaves, held opposite each other, one just above the other, at base of stem. Variable in shape, from oval to narrower and more strap-shaped, tapering at base to a whitish, winged stalk. 1–5 small, lanceolate, bract-like leaves higher on stem. **SPIKE** 5–25 flowers (occasionally more) form a compact cylindrical spike in moorland plants, more widely spaced in woodland plants. **BRACT** Green, narrow, pointed and slightly shorter than ovary. **OVARY** Pale green, long, narrow, clearly ribbed and twisted and also curled into a C-shape to hold the flowers pointing outwards and slightly downwards. **FLOWER** Sepals white, washed greenish towards tip and bluntly lanceolate in shape; upper sepal slightly broader and more triangular and may be bent upwards at tip. Petals creamy, variably washed greenish, smaller, narrower and more strap-shaped than sepals; upper sepal and petals form a loose hood over the column, lateral sepals held spreading and slightly drooped. **LIP** Creamy, sometimes greener towards tip, narrow, strap-shaped, 6–12mm long; projects forwards and downwards. Spur long and slender, 1mm wide x 13–23mm long (–27mm), sometimes slightly curved; may be washed green. **COLUMN** Pollinia whitish, *c.* 2mm tall, parallel, 1mm apart, at front of column. **SCENT** Heavy and sweet, sometimes likened to carnations, especially pungent at night. **SUBSPECIES** None. **VARIATION** In both Greater and Lesser Butterfly Orchids plants growing in the open tend to be shorter and more compact than those in shaded woodland localities. In Lesser Butterfly Orchid the 'heathland' form has leaves that are egg-shaped and a relatively dense flower spike whereas the 'woodland' form has narrower and less pointed, more tongue-shaped leaves. Intermediates occur and the differences between the two forms may be caused solely by the different environments. **Var. *trifolia*** has three main leaves rather than two. It is not rare. **Var. *quadrifolia*** has four leaves. **HYBRIDS** ***P.* x *hybrida***, the hybrid with Greater Butterfly Orchid, has been reported from scattered localities where the two species occur together but is rare; the hybrid is intermediate in the positioning of the pollinia and in spur length, and is fertile.

SMALL WHITE ORCHID *Pseudorchis albida*

IDENTIFICATION

A boreal species, rather local in Scotland and rare in N England, Wales and Ireland. Height 8–20cm (–40cm). The combination of small stature, dense spikes of creamy-white flowers and a deeply 3-lobed lip is distinctive. Flowers bell-shaped and very small, just 2–4mm across (smaller even than the ovary). The spike is carried on a long stem with a cluster of shiny green leaves at the base. **SIMILAR SPECIES** Creeping Lady's-tresses is superficially similar but flowers later in the year, in pinewoods (very rarely on moorland). Its flowers similarly have glandular hairs but its lip is rather different, spout-shaped rather than 3-lobed.

Irish Lady's-tresses flowers even later in the summer and has larger flowers arranged in three columns around the spike. Its lip is also formed into a spout.

Dense-flowered Orchid is superficially similar and the flowering periods may just overlap, but its lip is very different, with the hood tightly closed.

White-flowered varieties of Pyramidal, spotted and fragrant orchids have been mistaken for Small White Orchid, but their flowers are larger and differ in many other details. **FLOWERING PERIOD** Late May–mid July, latest at higher altitudes in the N and earliest in Ireland but generally at its best around mid June. The flowers tend to wither quickly.

HABITAT

Rough grassland on poor, well-drained soils, both mildly acidic and base-rich: hill pastures, hay meadows, road verges, banks, streamsides and grassy ledges. Will grow in the partial shade of shrubs and bushes and sometimes found on recently burnt moorland among short Heather, disappearing again once the Heather regenerates to form a closed community. Very rarely, found in oak woodland on acid soils and on stabilised coastal dunes. Recorded as high as 500m in Cumbria and 550m in Argyll.

POLLINATION & REPRODUCTION

The flowers produce nectar and are visited by butterflies, day-flying moths and solitary bees. The specific pollinator has not been identified, but the narrow entrance to the spur suggests that it may be butterflies. Some self-pollination also occurs, as the pollinia eventually fall onto the stigma if an insect has not removed them. Seed may be set by over 90% of flowers.

DEVELOPMENT & GROWTH

The roots and the slender tips to the tubers have a heavy fungal infection. The first aerial stem is reported to appear four years after germination.

STATUS & CONSERVATION

Listed as Vulnerable. Populations tend to be small and scattered and even in Scotland many vice-counties have just 1–2 sites for the species. Commonest and most widespread in N and W Scotland, including the Inner Hebrides and Orkney; in the

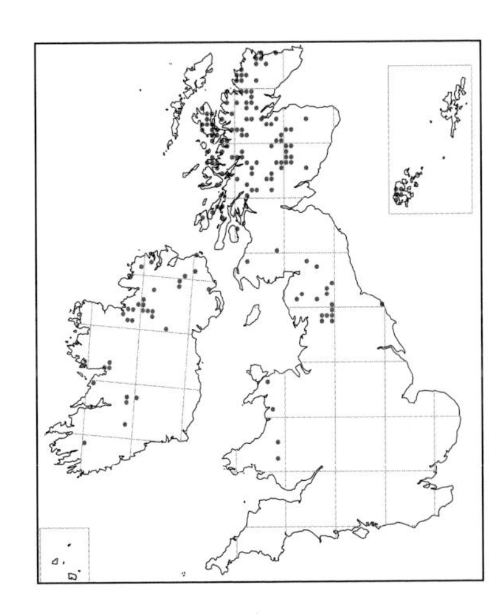

Colonies were lost throughout the 20th century due to habitat destruction, forestry, agricultural improvement and overgrazing. As with many grassland orchids it requires both a short sward and areas of bare soil so that seedlings can become established. It therefore needs a certain level of grazing to thrive; too little and it is swamped by coarse grasses and scrub, too much and it can never flower and is eventually eliminated. Climate change may also be having an impact as, in common with many other northern plants, it has declined drastically to the south of its range and appears to be withdrawing northwards.

DESCRIPTION

Underground The aerial stem grows from paired tubers that taper gradually to long, pointed tips and are often deeply divided into several long 'fingers' that diverge widely. There are also long, fleshy roots

remainder of Scotland has almost vanished from the area south of a line from the Clyde to Aberdeen. Similarly much reduced in N England, where now found rather rarely in Cumbria (a former stronghold that still holds the bulk of the English populations, although most colonies comprise just a handful of plants), as well as Northumberland, Co. Durham and mid-W and NW Yorks. In Wales numbers have collapsed in recent years and now found at a few scattered sites in Brecon, Merioneth, Denbigh and Caernarvon. In Ireland it is similarly very scattered, with the main concentration in the NW and in Northern Ireland.

Small White Orchid used to occur sparingly in S England, but is now extinct. The stronghold was East and West Sussex, where present until at least 1913, with three localities in West Sussex and ten in East Sussex, notably Ashdown Forest. In total, Small White Orchid has been lost from 66% of the historical range in Britain and 70% of the historical range in Ireland.

that lie horizontally close to soil's surface. **STEM** Greenish, slightly angled towards tip, with 2–3 whitish or brownish sheaths at the base. **LEAVES** 4–6 shiny, green, oval to oval-lanceolate, keeled sheathing leaves at base of stem with 1–2 narrower and more bract-like leaves above them. **SPIKE** Dense, cylindrical and often rather one-sided, with 20–40 flowers (exceptionally as few as 10 or as many as 70). **BRACT** Green, lanceolate with a pointed tip, as long as or just longer than ovary, which it clasps. **OVARY** Green, slightly twisted, with three obvious ridges, strongly curled over towards tip so that the flower faces more-or-less downwards. **FLOWER** Sepals whitish or creamy, petals, lip and spur whitish washed more greenish or yellowish. Sepals elliptical and blunt-tipped, forming a loose hood that encloses the similarly shaped petals and column. **LIP** Short, broader than long and deeply 3-lobed. Central lobe triangular, usually longer, wider and blunter than side-lobes, which are more lanceolate in shape. Spur short (2–3mm), tubular or sack-shaped, blunt-tipped and downcurved, containing abundant nectar. **SCENT** Delicate, recalling vanilla. **SUBSPECIES** British plants belong to the nominate subspecies, ***P. a. albida.*** **VARIATION** Nominate *albida* is divided into two varieties. **Var. *albida*** on more acid soils, with the lateral lobes of the lip clearly shorter than the central lobe, and **var. *tricuspis*,** usually on calcareous soils, with the lateral lobes almost as long as the central lobe. The distribution and abundance of the two varieties in the British Isles has not been studied. **INTER-GENERIC HYBRIDS** The hybrid with Heath Fragrant Orchid has been recorded from several places in Yorks and Scotland, and it is fairly frequent in NW Scotland (unnamed; the name X *Pseudadenia schweinfurthii* probably relates to the hybrid with Chalk Fragrant Orchid, which has not yet been recorded in Britain). **X *Pseudorhiza bruniana*,** the hybrid with Heath Spotted Orchid, was recorded from Orkney in 1977 and Skye in 1994.

CHALK FRAGRANT ORCHID

Gymnadenia conopsea

IDENTIFICATION

Very locally abundant throughout England and Wales but rather rare in Scotland and Ireland. Height 15–30cm (10–40cm, rarely to 60cm). Fragrant orchids are easy to separate from other orchids. The pink flowers are held in a tall, narrow, spire-like spike. Each flower has two 'wings' (the lateral sepals) that are held roughly horizontally, a flat, 3-lobed lip and a long, slender, down-curved spur.

SIMILAR SPECIES Pyramidal Orchid has flowers of a similar colour, size and shape and also has a long slender spur. Pyramidal Orchid has, however, an obviously shorter, more conical and more closely-packed flower spike, deeper pink flowers with a more markedly 3-lobed lip that has two diagnostic parallel raised ridges at the base. It also flowers later in the summer, and there is usually relatively little overlap.

The three fragrant orchids are difficult to separate, although usually found in distinct habitats. In Chalk Fragrant Orchid the flowers are medium-sized, the lip is about as wide as long, lacks 'shoulders' and is distinctly 3-lobed, with the central lobe the longest. The 'wings' (lateral sepals) are angled downwards and are narrow, pointed and parallel-sided. See p.145.

FLOWERING PERIOD Late May (sometimes mid May)–late July, but mostly June.

HABITAT

Dry, species-rich grassland on calcareous soils, mostly chalk downland in the S and limestone pastures in N England, but sometimes also on stabilised dunes, road verges, railway banks and in old quarries on suitable soils. Very occasionally occurs in base-rich fens (together with Marsh Fragrant Orchid) or on alkaline *Leblanc* waste in Lancashire. Recorded up to 365m above sea level (Northumberland).

POLLINATION & REPRODUCTION

Nectar is produced in the bottom of the long spur, and only insects with a long proboscis can reach this. Pollinators include butterflies and both day- and night-flying moths. Night-flying moths may, however, be the most important pollinators; the flower's scent becomes more pungent towards dusk, and, with its white flowers, var. *albiflora* may only be attractive to nocturnal moths. As the insect advances to sip nectar from the spur, the pollinia are fixed by their sticky viscidia (which lie just above the mouth of the spur) to its proboscis. The pollinia are then carried to another flower or another plant, having in the meantime swung forward into a position ready to make contact with the stigma. Pollination is very efficient and seed is set in large quantities.

Vegetative reproduction may occur via the formation of additional tubers at the base of the stem. These will go on to form separate plants as the connecting stem dies off in the autumn.

DEVELOPMENT & GROWTH

The roots and the root-like tips of the tubers are infected with fungi. Early

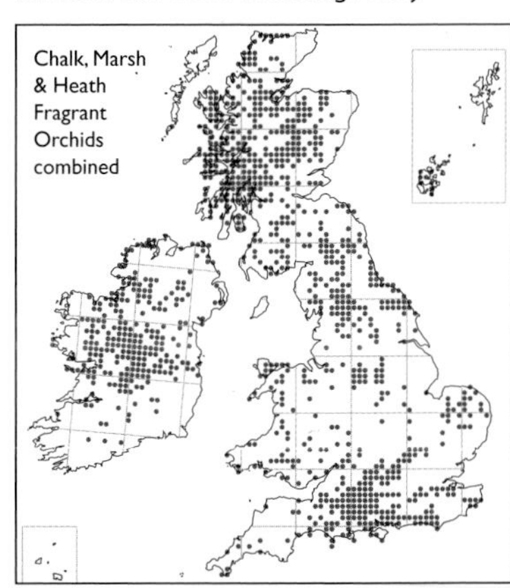

observations suggested that seeds probably germinate in spring. The resulting protocorm produces roots in its second autumn, the first leaves in its third spring (i.e. when two years old) and the first tuber the following season. Given the dry habitats in which it often grows, however, it seems unlikely that protocorms could survive for over two years and the period prior to the development of the first tuber may be much shorter. The interval between germination and flowering is usually about five years but may be as short as three.

STATUS & CONSERVATION

Common or even abundant at suitable sites, but declined significantly in the 19th and 20th centuries due to the 'improvement' of downs and pastures or their conversion to arable. Even where habitat still remains, like all grassland orchids this species requires a specific level of grazing; too little and scrub invades, too much and the diversity of species is lost

Since 1997 the 'Fragrant Orchid' has been separated into three distinct species, Common, Marsh and Heath Fragrant Orchids. Long treated as separated varieties or subspecies, molecular and genetic evidence has shown that they merit specific status, even though they can be hard to identify with certainty. Recording of the three fragrant orchids is still incomplete and it is hard to identify trends for the individual species. But, taken together, 'Fragrant Orchid' has vanished from 39.5% of the total historical range in Britain and 30.5% in Ireland.

DESCRIPTION

Underground The aerial shoot grows from a pair of flattened, deeply divided tapering tubers. Roots develop in autumn and penetrate the upper layers of soil. **Stem** Green, becoming more purplish towards flower spike. **Leaves** Mid green. 3–5 narrow, strap-shaped basal leaves, keeled with pointed tips, held loosely erect, grade into 2–3 narrow, lanceolate, bract-like leaves higher on stem. **SPIKE** More-or-less cylindrical, with 20–50 flowers, rarely more. Moderately dense, but becomes looser as more flowers open. **BRACT** Green, sometimes tinged purple, strap-shaped, narrowing around mid point to a finer pointed tip, about as long as ovary. **OVARY** Green, variably washed purple, long, narrow, prominently 3-ribbed and twisted. **FLOWER** Pink with a hint of purple, varying in its exact shade. Sepals elongated-oval, lateral sepals slightly irregular but with upper and lower margins rolled backwards so they appear parallel-sided with a short pointed tip; held *c.* 30° below the horizontal and pressed backwards. Petals a little shorter, more oval-triangular and asymmetrical, with one side squared off. Upper sepal and petals form a hood over the column. **LIP** Flat, about as wide as long, with three well-developed, rounded terminal lobes; central lobe longer than lateral lobes and often broader. Spur a darker, more purplish-pink, very slender, downcurved and rather long, about twice the length of the ovary, filled with nectar. **SCENT** Strong and sickly-sweet with rancid overtones. **SUBSPECIES** None. **VARIATION** **Var. *albiflora*** has white flowers and is fairly frequent. **Var. *crenulata*** has broad lateral lobes to the lip, narrowly serrated at the edges. Rare. **HYBRIDS** The three fragrant orchids may hybridise (although they seldom grow together), but confirming hybrids would be very difficult. **INTER-GENERIC HYBRIDS** X ***Gymnanacamptis anacamptis***, the hybrid with Pyramidal Orchid, has been recorded once each in Hants and Co. Durham; **X *Gymnaglossum jacksonii***, the hybrid with Frog Orchid, has been noted rarely but widely; **X *Dactylodenia heinzeliana***, the hybrid with Common Spotted Orchid, is uncommon but has been found widely in England and Wales. **X *Dactylodenia legrandiana***; the hybrid with Heath Spotted Orchid, has been found widely scattered in Britain and Ireland; **X *Dactylodenia wintoni***, the hybrid with Southern Marsh Orchid, has been found in Devon and Hants.

MARSH FRAGRANT ORCHID

Gymnadenia densiflora

IDENTIFICATION

Recorded from scattered localities throughout the Britain and Ireland but rather scarce. Height 30–60cm (–90cm); plants in chalk grassland are the smallest.

With spire-like spikes of pink flowers, each with a distinctly 3-lobed lip and a long, slender, curved spur, this species is easy to identify as one of the three fragrant orchids. **Similar species** The fragrant orchids can be hard to separate from each other, but Marsh Fragrant is usually distinctive in its habitat, late-flowering and is a tall, robust plant with numerous broad basal leaves that are usually held noticeably erect and a good-sized spike of relatively large, dark pink flowers. The lip is broader than long, prominently lobed and has distinct 'shoulders'. The specific name *densiflora* refers to the densely packed flower spike, although the spike is not necessarily more crowded than Chalk Fragrant Orchid. See p.145. **Flowering period** Late June–mid August.

HABITAT

Usually fens and meadows flushed with calcareous, base-rich water, also dune slacks and slumped, flushed, clay cliffs. Frequently found with Marsh Helleborine. Very occasionally occurs on chalk grassland on north-facing slopes. Recorded up to 310m above sea level (Banff).

POLLINATION & REPRODUCTION

No specific information but the differences in the size and shape of the flower, notably the long spur in Marsh Fragrant Orchid, suggest that a different suite of pollinators is involved for each of the three fragrant orchids. See also Chalk Fragrant Orchid.

DEVELOPMENT & GROWTH

No specific information.

STATUS & CONSERVATION

Originally described as a distinct species in 1806, but quickly reduced to 'var. *densiflora*' of 'Fragrant Orchid' in almost every field guide (if it was mentioned at all). Recently, however, genetic evidence has shown that it merits full specific status, but poorly known and listed as 'Data Deficient' because it has only recently been recorded systematically and identification criteria are still being developed.

Records are very scattered across Britain and Ireland, north to Orkney and including a few from the Inner and Outer Hebrides, but largely absent from Cornwall and Devon, most of mainland Scotland (away from the N and NW coasts and fringes of the Cairngorm massif) and the Republic of Ireland. There is otherwise very little pattern to the distribution, perhaps reflecting the scattered nature of suitable habitats, or the uncertainty still surrounding identification (no map).

Marsh Fragrant Orchid has undoubtedly undergone a serious decline in much of Britain. The direct drainage and destruction of fens and marshes has caused some losses. However, subtler effects, including eutrophication and the lowering of water tables, have also caused declines which reserve or SSSI status cannot protect against. The few chalk grassland populations are also subject to losses due to 'improvement' or abandonment.

DESCRIPTION

STEM Green, washed purple towards tip. **LEAVES** A rosette of several erect, lanceolate and relatively broad leaves at base of stem and several smaller, narrow and more bract-like leaves higher up. **SPIKE** Tall and narrow, with up to 100 flowers. **BRACT** Green, washed purple (especially towards edges), roughly length of ovary and narrowly oval, tapering to a point. **OVARY** Green, variably and sometimes heavily washed purple; long, narrow, curved, prominently 3-ribbed and twisted. **FLOWER** Deep pink, variably tinged purple and often whiter around base of lip. Sepals elongated-oval; lateral sepals held horizontally with upper and lower margins rolled back – they appear parallel-sided with a blunt tip. Petals a little shorter, more oval-triangular; upper sepal and petals form a hood over the column. **LIP** Flat, broader than long with distinct 'shoulders' and three well-developed rounded lobes, the side lobes larger than the central lobe. Spur relatively dark, very slender, downcurved and long. **SCENT** Spicy sweet, recalling cloves, lacking the rancid overtones of Chalk Fragrant Orchid. **SUBSPECIES** None. **VARIATION** The dune-slack population at Kenfig in S Wales has recently been assigned to **var. *friesica***, described from Dutch plants and said to be intermediate in appearance between Marsh and Heath Fragrant Orchids. The validity here of var. *friesica* and its relationship to the three species of fragrant orchid has, however, yet to be clarified. **INTER-GENERIC HYBRIDS** The hybrid with Common Spotted Orchid is uncommon, but probably occurs throughout the range. The hybrid with Southern Marsh Orchid has been recorded in S England and S Wales. Neither has a formal name.

HEATH FRAGRANT ORCHID

Gymnadenia borealis

IDENTIFICATION

The commonest fragrant orchid in N and W Britain, with scattered outposts in S England. Height:10–25cm (–30cm). A relatively small, delicate, few-flowered cousin of Chalk Fragrant Orchid. The conical spikes of pink flowers, each with a vaguely tri-lobed lip and long slender spur, identify it as one of the three fragrant orchids. **SIMILAR SPECIES** Usually separated from Chalk and Marsh Fragrant Orchids by habitat and its relatively small flowers with an obscurely-lobed lip that is longer than wide with small side-lobes and a longer central lobe; the flower spike is often rather flat-topped. Caution is required, however; see p.145. **FLOWERING PERIOD** June–August in Scotland and Ireland and early June–early July in the New Forest (around the same time as Chalk Fragrant Orchid or just a little later); anomalously, the population on chalk grassland in Sussex flowers late, July–early August.

HABITAT

In the N and W found on unimproved hill pastures, roadside verges and grassy moorland, often in areas flushed with ground water. In such places it grows on tussocks of grass and heathers, sometimes with Lesser Butterfly Orchid and Heath Spotted Orchid. Also found on machair, dunes and in old quarries. Tolerant of a wider range of pH than Chalk and Marsh Fragrant Orchids and will grow on soils that are both mildly acidic and alkaline, on sands, limestones and clays. In the New Forest and Ashdown Forest found in 'marl bogs' (flushes and hollows on grassy heaths on base-rich clays). Very rarely, it has been recorded from chalk grassland, as near Lewes in Sussex. Occurs up to 610m above sea level (Perthshire).

POLLINATION & REPRODUCTION

No specific information. See Chalk Fragrant Orchid.

DEVELOPMENT & GROWTH

No specific information.

STATUS & CONSERVATION

The most poorly known of the three fragrant orchids, only really brought to the attention of botanists in 1988 when it was treated in the Botanical Society of Britain & Ireland's *Plant Crib*. Its history and conservation is obscure, but there has undoubtedly been a considerable decline, with overgrazing of upland grasslands being a particular problem.

Widespread and locally common in Scotland, especially in the highlands, and including the Inner Hebrides, Orkney and Shetland (Unst). Much more scattered in S Scotland, and in N England found in the Lakes, the Pennines and the North York Moors, with an outpost in the Peak District in Derbyshire. In England otherwise known from Shropshire, Cornwall (including the Lizard Peninsula and Bodmin Moor), Devon, Dorset, the New Forest in Hants (*c.* 12 relatively small populations) and E and W Sussex (Ashdown Forest and three sites on the South Downs). Scattered records from W Wales and NW and N Ireland, where mostly coastal, but very few records from the Republic of Ireland. The distribution clearly has a northern and western bias, but the species is still poorly-documented (no map).

DESCRIPTION

STEM Green, becoming more purplish-brown towards tip. **LEAVES** 3–5 narrow, strap-shaped basal leaves, keeled, pointed, held loosely erect, often in two ranks, grade into 2 –3 narrower and more lanceolate bract-like leaves higher on stem.

SPIKE Fairly open, more-or-less cylindrical, although often slightly irregular in shape, with 20–30 or more flowers. **BRACT** Green, tinged purple (especially on edges), strap-shaped, narrowing abruptly around mid-point to a finer pointed tip; *c.* 1.5x length of ovary. **OVARY** Green, variably washed purple, long, narrow, 3-ribbed and twisted. **FLOWER** Dark pink to lilac. Sepals oval or drop-shaped; lateral sepals held a little below the horizontal and pressed backwards, margins rolled back to give a pointed, oval-lanceolate shape. Petals shorter, more oval-triangular but less regular in shape. Upper sepal and petals form a hood over the column. **LIP** Flat, longer than wide, lobed at tip, side-lobes shorter than central lobe; lobes usually poorly developed with notches between them reduced or absent. Spur darker, more purplish-pink, very slender, downcurved and rather long, about twice as long as the ovary and filled with nectar.

SCENT Powerful and sweet, recalling cloves. **SUBSPECIES** None. **VARIATION** Var. ***albiflora*** has white flowers and is fairly frequent. Plants with creamy-yellow flowers have been recorded very rarely in Co. Donegal. **INTER-GENERIC HYBRIDS** The hybrid with Small White Orchid has been recorded from Yorkshire and Scotland and the hybrid with Early Marsh Orchid has been found once, in W Cornwall; neither has a formal name. **X *Dactylodenia varia***, the hybrid with Northern Marsh Orchid, has been found in N England, Scotland and Co. Down. **X *Dactylodenia st-quintinii***, the hybrid with Common Spotted Orchid, is uncommon but found widely in N England and Scotland. **X *Dactylodenia evansii***, the hybrid with Heath Spotted Orchid, has been found in N England and Scotland.

Identification of Fragrant Orchids

It is best to examine several plants in a population to gain an overall impression. The criteria used to separate the three fragrant orchids are still being developed, and some plants or even whole populations may not yet be identifiable.

	Chalk Fragrant Orchid	Marsh Fragrant Orchid	Heath Fragrant Orchid
Width of florets*	10–11mm (7–13mm)	11–13mm (10–14.5mm)	8–10mm (7–12mm)
Lip: width	5.5–6.5cm (4.5–7mm)	6.5–7mm (5.5–8mm)	3.5–4mm (3–5mm)
Lip: length	5–6mm (4–6.5mm)	3.5–4mm (3–4.5mm)	4–4.5mm (3.5–5mm)
Lip: shape	fractionally broader than long	much broader than long	longer than wide
Lip: lobes	distinctly 3-lobed, central lobe longest	distinctly 3-lobed, side-lobes largest	obscurely lobed
Lip: shoulders**	absent	distinct	narrow
Lateral sepals: width	c. 1mm	c. 1mm	c. 2mm
Lateral sepals: length	c. 5–6mm	c. 6–7mm	c. 4–5mm
Lateral sepals: shape	linear, pointed at tip	linear, blunt at tip	oval-lanceolate, pointed at tip
Lateral sepals: position	deflexed at c. 30°	horizontal	deflexed
Length of spur	12–14mm (11–17mm)	14–16mm (13–17mm)	11–14mm (8–15mm)
Height	20–40cm, sometimes more	30–60cm, less in dry habitats	15–25cm, rarely more
Flowering period	early June-mid July	early July–mid August	June–August
Habitat	chalk & limestone grassland	fens; rarely chalk grassland on north-facing slopes	mildly acidic to base-rich grassland; base-rich heathland flushes; very rarely chalk grassland

Largely after Francis Rose in *The Plant Crib* (1998).

* The width of the floret is measured from tip to tip of the lateral sepals

** 'Shoulders' – abrupt angles at the base of the lip as it narrows towards the mouth of the spur.

FROG ORCHID *Coeloglossum viride*

IDENTIFICATION

Widespread but rather local, and rare in E England. Height 5–15cm (4–25cm, exceptionally to 45cm). The small size, generally greenish or reddish-purple flowers, tight hood and rather plain, strap-shaped lip are distinctive. **Similar species** Of the vaguely similar species, Man Orchid has long, narrow arms and legs on its lip and Common Twayblade has a smaller flower, again with arms and legs. **Flowering period** Very variable. Mid May–early August (–early September); some populations are early, some much later, perhaps especially those on chalk grassland in S England. Once pollinated, the flowers persist for a long time, although the lip will have withered.

HABITAT

In S Britain largely confined to well-drained short grassland on chalk or limestone, especially the slopes of ancient earthworks, abandoned quarries, old chalk and lime pits and spoil heaps. Tolerant of grazing and trampling but cannot compete with rank vegetation. Formerly also found in damp or wet permanent pastures and meadows, but this habitat has almost entirely vanished in S England. In the N and W found in a wider range of short-grass habitats, often damp, on both calcareous and neutral soils, including limestone pavements, rocky ledges, road verges, railway embankments, upland flushes, mountain pastures, coastal grassland, machair and dune slacks. Occurs from sea level to 915m (Angus).

POLLINATION & REPRODUCTION

Nectar is secreted into the short spur and a variety of insects visit the flowers, including small beetles and wasps, especially ichneumons. These alight on the lip and are directed by its central ridge to either side; as they approach the spur, the pollinia are stuck to their head by the viscidia which lie on either side of the spur. Once on the insect's head it takes *c.* 20 minutes for the pollinia to rotate forward to be in position to make contact with the stigma of the next plant to be visited. Self-pollination is also reported to occur. Seed-set is variable, with the capsules maturing rapidly and containing around 1,250–5,000 seeds. Only occasionally reproduces vegetatively.

DEVELOPMENT & GROWTH

The period between germination and flowering is short. The first green leaves appear 1–3 years after seed has germinated and a flower spike is usually produced in that first year above ground. Frog Orchid is short-lived with a rapid turnover of the population. Many plants are monocarpic and die after just one year above ground, although some may live and flower for at least seven years. In a study in Holland the 'half-life' of a population averaged 1.5 years and varied from 1 to 2.4 years (see p.251). Occasionally, plants may be dormant underground but not for more than one year.

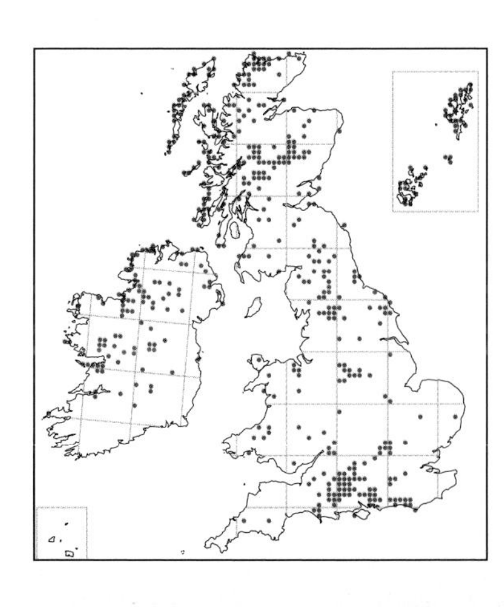

STATUS & CONSERVATION

Vulnerable. Widely distributed, including the Inner and Outer Hebrides, Orkney and Shetland, but has declined severely in much of lowland Britain. In S England it is now almost confined to the chalk districts of Hampshire, NE Dorset and Wiltshire, extending very locally into the Berkshire Downs, Chilterns and the South Downs in Sussex. In the Midlands it is more-or-less restricted to the Peak District. It is very local in Wales, lowland Scotland and the southern half of Ireland but commoner and more widespread in N England, upland Scotland and the northern half of Ireland.

Has declined throughout almost the entire range in the British Isles, with the biggest losses in England, especially the Midlands and E Anglia, where the ploughing-up of old pastures on neutral or chalky boulder-clay soils appears to have been a major factor, as well as the development of scrub and rank vegetation on what had been short grassland.

DESCRIPTION

Underground The aerial stem grows from a pair of tubers that are forked into 3–4 finger-like lobes. **Stem** Green, on some plants washed reddish-purple towards flowers, with 1–2 brown, leafless sheaths at extreme base. **Leaves** 3–5 dark green, strap-shaped sheathing leaves, sometimes rather broad, the two lowest the largest and bluntest; held at *c.* 30°–45° above the horizontal. Higher on stem several narrower and more lanceolate non-sheathing leaves. The rosette appears in autumn and overwinters, and there may be a relatively large number of non-flowering plants; these have fewer leaves.

Spike Rather loose, with 5–25 (2–50) flowers, often irregular in shape; ovaries variably twisted and flowers therefore pointing in different directions.

Bract Green, variably washed reddish-purple, lanceolate; lower bracts may be rather longer than flowers (up to a maximum of 2x flower's length), but towards tip of spike bracts shorter.

Ovary Green, variably washed reddish-purple, spindle-shaped, 6-ribbed and twisted. **Flower** Green or yellowish-green, variably washed brown or reddish-purple, in some plants whole flower may be reddish-purple. Sepals dark green variably washed reddish-purple, roughly oval to oval-triangular; upper sepal a little smaller than lateral sepals. Petals pale green, washed reddish-purple around edges and strap-shaped, much smaller and narrower than sepals. Sepals and petals form a tight hood over column, petals often showing as much paler slots

between sepals. **LIP** Usually paler and greener than rest of flower and even paler and yellower towards base, with reddish tones usually confined to tip and edges. Tongue-shaped, may broaden slightly towards tip, with two small terminal lobes that have a notch between them and a third smaller lobe within this notch; a thickened ridge down the centre of lip; nectar is secreted from two hollows formed by raised and curved margins of lip at its base. Lip may hang downwards or be folded backwards below ovary. Spur very short (1–2mm), almost hemispherical in shape and rather colourless. **COLUMN** Anthers washed purple, pollinia club-shaped, pale yellowish with a very lobed surface. **SCENT** Faintly honey-scented. **SUBSPECIES** None. **VARIATION** Averages shorter, stouter and more reddish-brown in the dry grassland habitats of S England and taller and greener in the N. Some plants may be entirely yellowish-green, including the flower spike. **Var. *longibracteatum*** has unusually long bracts and is rather taller and more robust. It has been found in N England. **INTER-GENERIC HYBRIDS** **X *Dactyloglossum mixtum***, the hybrid with Common Spotted Orchid, is rare but has been found at widely scattered localities. **X *Dactyloglossum conigerum***, the hybrid with Heath Spotted Orchid, has been found rarely. **X *Dactyloglossum viridellum***, the hybrid with Northern Marsh Orchid, has been recorded from Co. Durham and the Inner and Outer Hebrides. **X *Gymnaglossum jacksonii***, the hybrid with Chalk Fragrant Orchid, has been noted sporadically but widely in England. Hybrids with Heath and Marsh Fragrant Orchids probably also occur.

WHAT'S IN A NAME?

Recent genetic research has suggested that the genus *Coeloglossum* is sufficiently similar to the genus *Dactylorhiza*, the spotted and marsh orchids, for Frog Orchid to be transferred to it as *Dactylorhiza viridis*. This close relationship is reflected in Frog Orchid's propensity to form hybrids with members of the *Dactylorhiza*. Frog Orchid differs from *Dactylorhiza* not only in general appearance, however, but also in having a just rudimentary spur and bursicle, by producing nectar, and by having the 'hood' of the flower made up from all the sepals and petals (in the spotted and marsh orchids the lateral sepals are held away from the hood). For these reasons the change in Frog Orchid's affiliation has not been universally accepted.

The English name 'Frog Orchid' has been in use since the 17th century but it is hard to see the 'frog' in a Frog Orchid; the two-lobed lip could resemble hind legs and the hood may look like a frog's body, but to anyone except the most imaginative any real resemblance to a frog is fanciful.

Spotted and marsh orchids: the genus *Dactylorhiza*

The spotted and marsh orchids, genus *Dactylorhiza*, pose the toughest identification challenge amongst British and Irish orchids. They are closely related and individually quite variable. Hybridisation is also relatively common, making identification even more difficult. If in doubt it is usually necessary to carefully study a selection of plants in order to reach an identification. A ruler is useful (to accurately measure the lip of the flower), as is a notebook to keep a permanent record.

lateral sepal
upper sepal
lateral sepal
petal
petal
mouth of the spur
side-lobe
side-lobe
notch ('sinus')
double loop markings
central lobe

The *Dactylorhiza* orchids are divided into two groups depending on the number of chromosomes they possess. The diploid group has a chromosome count of 2n = 40. This includes Early Marsh Orchid and Common Spotted Orchid. The tetraploid marsh orchids have a chromosome count of 2n = 80. This group includes Southern, Pugsley's, Northern and Irish Marsh Orchids. The chromosome count is significant for two reasons. The first is hybridisation, as it has some influence on the fertility of hybrids. Second, the tetraploid marsh orchids are thought to have evolved from a cross between the ancestors of two members of the diploid group, Early Marsh Orchid and Common Spotted Orchid, followed by a doubling of the chromosomes (in which 2n = 40 became 2n = 80). This allowed the hybrid to become reproductively isolated from its parents. This hybridisation event occurred several times and with slightly differing parents, producing the closely similar species we see today.

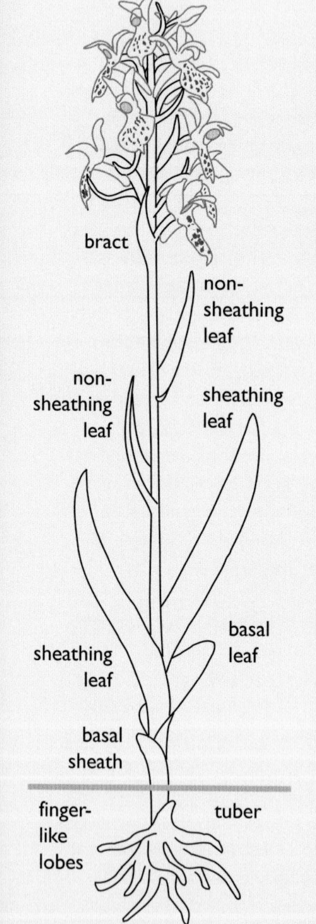

Pollination

Dactylorhiza are cross-pollinated by insects, especially bees, but they produce no nectar and visiting insects receive no reward. The brightly coloured flowers, especially en masse, deceive the insects, and the deception may be helped by the variation between individuals (perhaps especially in Common Spotted Orchid), which undermine the insects' ability to 'learn' the deception and avoid the flowers.

Growth pattern

The aerial stem grows from a pair of tubers that are flattened and divided into 2–5 finger-like lobes that taper to a fine point. These may be long in dry habitats, extending well down into the soil, but in wet areas may bend upwards towards the surface, perhaps to avoid being waterlogged. There are also several long, fleshy roots growing near the surface of the soil.

DACTYLORHIZA HYBRIDS

Hybrids in this genus are common. They are mostly found where two habitats (and thus two species) meet or in disturbed habitats, particularly those where large populations of several species have developed. If hybrids are sterile they are likely to be found either singly or in very small numbers, whereas even partial fertility can result in large numbers of hybrids forming – first generation (F_1) hybrids can 'back-cross' with the parent species and with each other to form a 'hybrid swarm' – eventually there may be no 'pure' plants left at the site. Hybrids can only be identified after careful study, and only with a good knowledge of the parent species. They will tend to be intermediate in most or all characters, and some first generation hybrids will show 'hybrid vigour' and may be larger than either parent – perhaps taller, or with more leaves, more flowers, or larger flowers with a larger lip.

***D. x transiens* Common Spotted x Heath Spotted** Widespread but rather scarce and hard to identify with certainty. Sterile.

***D. x kernerorum* Common Spotted x Early Marsh** Scattered throughout. Sterile.

***D. x grandis* Common Spotted x Southern Marsh** Found not only where both species occur together but also in the absence of one or other parent. Partial fertile, allowing back-crossing and the creation of long-lived hybrid swarms. Probably the commonest hybrid orchid.

***D. x venusta* Common Spotted x Northern Marsh** Occurs throughout the range of Northern Marsh and can be common. Partially fertile, and some evidence that back-crossing can allow the creation of hybrid swarms (photo p.178).

Common Spotted x Pugsley's Marsh Scarce. Recorded from Anglesey, Yorks and Ireland. Sterile. (No formal name.)

Common Spotted x Irish Marsh Rare, recorded in Co. Limerick and Co. Clare. (No formal name.)

***D. x carnea* Heath Spotted x Early Marsh** (mostly subsp. *pulchella*). Scattered in Britain and Ireland but scarce. Sterile.

***D. x hallii* Heath Spotted x Southern Marsh** Infrequent as the parent species favour different habitats. Fertile and can create hybrid swarms.

***D. x formosa* Heath Spotted x Northern Marsh** Occurs throughout the range of Northern Marsh. Fertile and can form hybrid swarms. Probably the commonest hybrid orchid in N Britain and Ireland.

Heath Spotted x Pugsley's Marsh Rare, with scattered records within the range of Pugsley's Marsh. Probably fertile. (No formal name.)

***D. x dinglensis* Heath Spotted x Irish Marsh** Found widely but under-recorded. Fairly fertile and capable of back-crossing to create hybrid swarms.

***D. x wintoni* Early Marsh x Southern Marsh** Scattered throughout the range of Southern Marsh but rare. Limited fertility, but some F_1 offspring may be very like Southern Marsh and hard to detect.

***D. x latirella* Early Marsh x Northern Marsh** Recorded throughout the range of Northern Marsh. Rare, probably sterile.

Early Marsh x Pugsley's Marsh Found in NW Wales, Yorks and Ireland. Rare, probably sterile. (No formal name.)

Early Marsh x Irish Marsh One record from Co. Clare. (No formal name.)

***D. x insignis* Southern Marsh x Northern Marsh** Scarce, but hard to identify. Probably fertile and likely to back-cross.

Southern Marsh x Pugsley's Marsh Fairly common in S English colonies of Pugsley's Marsh. Fertile. (The S England colonies of Pugsley's Marsh may be better treated as Southern Marsh, in which case this hybrid combination would no longer exist. See p.173; no formal name.)

Northern Marsh x Pugsley's Marsh A few, scattered records. (No formal name.)

Northern Marsh x Hebridean Marsh 'Intermediate' plants have been found but may be merely variant Northern Marsh.

Common Spotted Orchid

Spike Often pyramidal or conical as the flowers begin to open, becoming longer and more cylindrical; 20–70 flowers, exceptionally 150.
Flower Colour Usually has a distinct ***lilac tone***, but often very pale, even white.
Lip Cut ***about halfway*** to base by deep wedge-shaped notches into three ***large lobes*** (central lobe at least half the width of the side-lobes and usually a little longer too).
Lip Markings Bold ***purplish*** dashes and broken lines, often forming a pattern of concentric double loops (occasionally just dots and dashes); but often faint.
Spur *Long and slender,* parallel-sided or slightly tapering and either straight or very slightly curved.
Lateral Sepals Usually ***spreading horizontally*** (but may be closer to 45°).
Leaf Markings Usually ***marked all over with solid dark spots***, often large and elongated into bars. Unspotted leaves infrequent.
Sheathing Leaves Crowded together at base of stem, the lowest relatively ***small, oval and blunt-tipped***; remainder relatively ***broad***.
Non-sheathing Leaves *2–6 (1–9).*
Stem Usually ***solid*** but may be hollow, especially in larger plants in wetland habitats.

Heath Spotted Orchid

Spike Usually rounded or pyramidal, even when flowers all open; 5–20 flowers, exceptionally 60.
Flower Colour Whitish to ***very pale lilac***, sometimes deep pinkish-lilac.
Lip Obviously ***broad***, roughly circular, cut by shallow, triangular notches into three ***very unequal lobes***; side-lobes very large, central lobe ***small, and rather tooth-like***.
Lip Markings Darker dots and dashes, usually ***rather fine*** **and forming a pattern of concentric loops or circles but sometimes bolder, forming clear double loops.**
Spur *Long and slender,* parallel-sided or slightly tapering and either straight or very slightly curved.
Lateral Sepals *Spreading horizontally* or drooping.
Leaf Markings Usually ***marked all over with small, rounded (or only slightly elongated), dark spots***. Unspotted leaves quite frequent.
Sheathing Leaves Long, ***narrow***, obviously keeled and pointed, crowded towards base of stem; lowest leaf ***not significantly shorter or blunter*** than remainder.
Non-sheathing Leaves *2–5 (1–7).*
Stem *Solid.*

Early Marsh Orchid

Spike Crowded; 10–70 flowers.
Flower Colour *White to very pale pink*, sometimes deeper, more reddish-pink, but several subspecies – deep red (dunes), purplish-pink (fens and acid bogs), creamy (fens - very rare) or white.
Lip *Small*, usually less than 8.5mm wide × 7mm long, slightly to moderately 3-lobed, ***appears narrow***, with the sides usually ***clearly and sharply folded downwards***, sometimes so much so the lip is almost folded in two.
Lip Markings Dark pink dots and dashes ***contained within two complete or near-complete loops***; edges ***unmarked***.
Spur *Squat and sack-like*: stout, slightly tapering and slightly to moderately decurved (rarely straight).
Lateral Sepals Usually held ***vertically over flower***.
Leaf Markings *No dark markings* except in 'Flecked Marsh Orchid' (NW Scotland and W Ireland), which can have bold spots.
Sheathing Leaves Often crowded towards base of stem and typically rather ***erect***, also ***broad***, strongly keeled and often ***hooded*** at tip.
Non-sheathing Leaves 1–2.
Stem Usually ***hollow***, conspicuously so in larger plants.

Southern Marsh Orchid

Spike Rather crowded, conical as first flowers open, becoming cylindrical later; 20–60 flowers, large plants may have over 100.
Flower Colour Purplish-pink, sometimes slightly drab or washed-out.
Lip Mostly over 10mm wide, ***roughly circular***, cut by ***shallow*** notches into three ***indistinct*** lobes, side-lobes rather broad and rounded, central lobe (if obvious) ***small and tooth-like***; lip flat, slightly dished or with the side-lobes turned downwards at base but often level or turn upwards towards tip, the lip therefore often appears ***drop-shaped*** or a well-rounded triangle in its natural position.
Lip Markings *Fine* dark spots and/or short dashes, ***usually concentrated in centre***; 'Leopard Marsh Orchid' has bold dark loops and dashes.
Spur *Squat and sack-like*: tapering slightly to a blunt tip; may be slightly downcurved.
Lateral Sepals Variable, from near vertical to nearly horizontal, but usually held at *c.* 45°.
Leaf Markings *None*, except in 'Leopard Marsh Orchid', which has ***bold dark ring-shaped spots*** (beware hybrids).
Sheathing Leaves Held erect at 45° or more. No marked keel but may be slightly hooded at tips.
Non-sheathing Leaves 1–3 (usually 2).
Stem Hollow.

Northern Marsh Orchid

Spike Dense, oval to cylindrical and often flat-topped when fully open; 10–40 flowers, exceptionally 80.
Flower Colour ***Rich, dark 'velvety' magenta*** with a ***distinct crimson tone*** when fresh.
Lip Relatively small. Variable but typically appears ***diamond-shaped*** with straight sides, especially to the base of the 'diamond' (actually shallowly 3-lobed with a small central lobe, held flat or dished, with the margins of the side lobes folded ***upwards*** to give the lip a diamond shape).
Lip Markings Heavy lustrous ***dark crimson*** lines, dots and swirls over much of lip in a concentric pattern.
Spur ***Squat and sack-like***: thick, conical, slightly downward pointing and shorter than ovary.
Lateral Sepals Held at c. 45°.
Leaf Markings ***Unspotted*** or sometimes with a few very small spots near the tip, except in parts of W Wales, N England and NW Scotland where var. *cambrensis* has ***dark spots all over leaves***.
Sheathing Leaves 4–8, slightly crowded towards base of stem, broad, hooded at the tip and held at up to 45° above the horizontal.
Non-sheathing Leaves 1–2 (0–4), usually just one.
Stem Slightly hollow.

Pugsley's Marsh Orchid

Spike Loose and rather irregular, the flowers ***all face in more-or-less the same direction***; flowers ***few*** 6–14 (2–18).
Flower Colour Purplish-pink, very variable in intensity.
Lip ***Relatively large***, averages ***at least 9mm wide***. Usually ***distinctly 3-lobed***, central lobe ***longer and rather narrower than side lobes***, sometimes ***turned downwards***; lip appears ***triangular or diamond-shaped***, narrowest at base.
Lip Markings Lines, dots and blotches in an irregular pattern ***extending more-or-less to edges***. 'Lapland Marsh Orchid' has very bold loops and lines.
Spur ***Sack-like***: long, straight and thick but tapering slightly to a blunt tip.
Lateral Sepals Held between 45° and vertical but is usually closer to latter.
Leaf Markings Unspotted or occasionally finely spotted (especially in Yorks and Ireland); 'Lapland Marsh Orchid' has ***boldly spots.***
Sheathing Leaves 2–4 (–5), ***narrow***, strap-shaped, ***well spaced*** and often held erect at c. 45°.
Non-sheathing Leaves 0–1 (–2); ***combined total of sheathing and non-sheathing leaves 3–5*** (excluding the short basal leaf),
Stem Slightly hollow, ***rather slender and relatively weak***.

Hebridean Marsh Orchid

Spike Short, compact and cylindrical; 5–20 flowers.
Flower Colour *Rich, deep purple-magenta*.
Lip *Broader than long, distinctly 3-lobed,* central lobe longer than side-lobes; flat or dished when flower first opens but the side-lobes may become slightly deflexed with age.
Lip Markings Heavy dark purple spots and dashes bounded within 1–2 concentric dark loops.
Spur *Squat and sack-like*: a little longer than ovary, thick, cylindrical and slightly tapering.
Lateral Sepals Held at 45° above the horizontal, or even higher.
Leaf Markings Usually moderately marked with rings or fine blotches (especially non-sheathing leaves), concentrated towards leaf tips and on more heavily marked plants they begin to coalesce; there is ***a continuous gradation from plants with unmarked green leaves to those with leaves that are solidly blackish-purple***; spotting is occasionally also present on the underside. Leaves finely rimmed with purple (especially uppermost), even when not spotted.
Sheathing Leaves 2–3 (–4), relatively broad, crowded towards base of stem, the lower leaves tending to be held nearer the horizontal than the vertical; the lowest leaf is often rather shorter than the rest.
Non-sheathing Leaves One.
Stem Hollow.

Irish Marsh Orchid

Spike Crowded; *c.* 20 flowers.
Flower Colour Pinkish-purple.
Lip *Broad*, and usually broader than long, broadest around the middle., with ***3 distinct, rounded lobes***; side-lobes variably reflexed, especially as the flower gets older; central lobe smaller, sometimes tooth-like and often projecting a little beyond side-lobes.
Lip Markings Heavily marked with dots and dashes, often enclosed within double loops (sometimes concentric double loops); there may also be markings outside the loops.
Spur *Squat and sack-like*: roughly conical, straight.
Lateral Sepals Usually held closer to horizontal than to vertical.
Leaf Markings *Spotted in c. 50% of plants*, spots usually concentrated on outer half of leaf and variable in shape; they may be rings or elongated into transverse bars.
Sheathing Leaves (3–) 4–6, fairly broad, usually grouped towards base of stem and held rather spreading, closer to the horizontal than the vertical.
Non-sheathing Leaves 2–3.
Stem Slightly hollow.

COMMON SPOTTED ORCHID

Dactylorhiza fuchsii

IDENTIFICATION

Widespread and common, often abundant. Height 15–30cm (7–70cm). Identification usually straightforward: leaves marked all over with *solid dark spots or bars* (the leaves may be heavily or lightly marked and plants with unspotted leaves are not uncommon); lip cut *around halfway to the base* into three *large lobes*; flower colour very variable, from almost white with very faint markings to lilac (whiter around the 'throat') with bold purple lines and loops; most plants are in between and the combination of ground colour and markings usually gives a distinctive *lilac tone* to the whole flower. Tends to be tall and pale-flowered in shaded sites, shorter and more compact in the open. Does best in damp or wet habitats; plants on dry chalk grassland can be very petite while those in exposed, windswept sites may be short and squat. **Similar species** See pp. 152–155. Early Purple Orchid has spotted leaves, but its lip is purple and a very different shape. It also flowers rather earlier in the spring.

Hybrids are common, particularly with Southern and Northern Marsh Orchids. Identification of hybrids requires a good knowledge of the range of variation of both species and even then is not always certain. First generation hybrids often show 'hybrid vigour' and may be obviously large, sometimes ridiculously so. Hybrids with marsh orchids often have the fat, conical spur of the marsh orchid parent rather than the thin spur of the spotted orchid but in subsequent generations back-crossed with the parent species, or in hybrid swarms, these distinctions will break down. **Flowering period** Mid May–early August: earliest in sunny, sheltered localities in the S (where generally best around mid June) and later in woodland sites and in the N.

HABITAT

Very varied. Dry grassland, from heavily grazed chalk downs to ranker and scrubbier sites, also thrives in damper conditions in wet meadows, the machair of NW Scotland, marshes, fens and dune slacks, including alkaline flushes in otherwise unsuitable moorland. Also woodland, usually better-lit areas along rides, clearings and edges, but can tolerate quite deep shade where it may persist in a non-flowering state.

Usually found on alkaline or neutral soils but can grow in slightly acidic conditions, e.g. less acid areas of heathland. An opportunist, it will colonise new sites, including man-made habitats such as abandoned gravel and chalk pits and industrial waste sites (e.g. fly ash tips and alkaline *Leblanc* waste). Often found on roadside verges and cuttings, railway embankments and sometimes lawns. Recorded up to 530m above sea level in Cumbria and 600m in Perthshire.

POLLINATION & REPRODUCTION

A wide variety of insects has been recorded carrying off pollinia from this species, but studies in Austria and Poland have

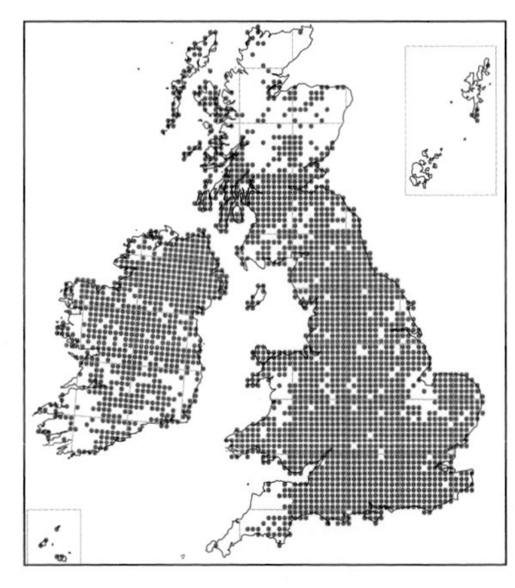

identified beetles, especially longhorn beetles, as the major pollinators. In common with other members of the genus *Dactylorhiza*, Common Spotted Orchid does not produce nectar and insects receive no reward for visiting the flowers. Whatever the mechanism, pollination is efficient and 50–90% of flowers produce seed. Notably, twice as many of the lower flowers, which open first, produce ripe capsules, compared with the upper flowers, which open later on. As the season progresses insect pollinators may become scarcer or lose interest as there is little or no reward, or there may be fewer resources available for seed production. Vegetative reproduction is also possible.

DEVELOPMENT & GROWTH

The period between germination and flowering in the wild is unknown but in cultivation can be as little as two years, although more often 4–5. Individual plants can flower for several years in a row or remain as a dormant non-flowering rosette if conditions become unsuitable.

STATUS & CONSERVATION

The most widespread British orchid and currently present in 67% of the total available 10km squares in both Britain and Ireland, including the Isle of Man, Inner and Outer Hebrides and Shetland. Absent, however, from most of Cornwall, much of N Devon and from large areas of N and NE Scotland. Formerly recorded from Orkney and the Isles of Scilly. In much of Britain and Ireland also the commonest orchid and often found in large numbers.

There has been some decline but Common Spotted Orchid is so versatile that it can usually 'hang-on' somewhere within any particular 10km square. The *New Atlas* figures are thus likely to conceal a significantly greater decline in the number of actual populations. Notably, much of the decline in Britain appears to be recent and it could be that a fall in the number of populations is increasingly expressing itself as a decline in the overall range. At least some of the losses are compensated for, however, by the colonisation of new areas, especially in recent man-made habitats.

Common Spotted Orchid is closely related to Heath Spotted Orchid and although they are clearly distinct in Britain, France and Scandinavia they are rather similar in central Europe and are considered to be the same species by many European authors. They are usually quoted as differing in the number of chromosomes, with Common Spotted Orchid having 2n=40 and Heath Spotted Orchid 2n=80, but various studies have given chromosome counts of 2n=40, 2n=60 and 2n=80 for both species (i.e. diploid, triploid and tetraploid). Even in Britain the differences between the two are not absolute. Common Spotted Orchid is generally found on calcareous or neutral soils and Heath Spotted Orchid on acid soils. However, there is some evidence that spotted orchids growing on intermediate, neutral or slightly acidic soils do tend to be intermediate in appearance.

DESCRIPTION

See also p.152. **Stem** Pale green, sometimes lightly washed purple towards tip; grooved. **Leaves** Green, usually marked all over with solid dark spots and blotches but sometimes unmarked; 3–6 (2–7) sheathing leaves, up to 4cm or even 5.5cm wide, lanceolate. **Bract** Green, sometimes washed purple, lanceolate and finely pointed; lowest longer than ovaries but higher on spike about equal in length. **Ovary** Cylindrical, 6-ribbed, twisted and green, often tinged violet. **Flower** Sepals oval-lanceolate, lateral sepals asymmetrical, marked with lines and spots; upper sepal and petals, which are a little shorter, form a hood over the column. **Lip** 8–12mm wide, a flattened oval cut into three roughly equal lobes; side-lobes with broad rounded tips, sometimes toothed, and roughly parallel sides; central lobe slightly narrower and more triangular but as long or longer. Lip held flat or with side-lobes lightly depressed. **Spur** 1/2–1x length of ovary. **Column** Pollinia pale brownish-pink to purple, sometimes yellowish. **Scent** Faint. **Subspecies** ***D. f. fuchsii*** is the commonest and by far the most widespread subspecies.

D. f. hebridensis* 'Hebridean Spotted Orchid** occurs in machair and similar coastal habitats in the Outer Hebrides, NW Scotland, Shetland and W Ireland. In some areas overlaps and mixes with subspecies *fuchsii*. Tends to flower relatively late. Characterised by dark flowers and a broad lip. Small and stocky, 8–20cm tall (6–40cm), spike densely packed, lip usually over 9.5mm wide (8–15mm), and flowers often deep rose-pink or reddish-purple (rarely white). Sometimes treated as a separate species by European authors but its distinctiveness has been exaggerated and, conversely, it is also sometimes treated as merely 'var. *hebridensis*'. **Variation** **Var. *alpina is a dark-flowered variant of subspecies *fuchsii*. Its dark flowers recall subspecies *hebridensis* but are smaller, the lip usually less than 9.5mm wide, with narrower side-lobes. Scotland, both inland and the machair of the Hebrides and NW coast (including some populations in the Inner Hebrides traditionally named '*hebridensis*'). Also inland sites in N England and possibly also Wales and W Ireland.

Var. *cornubiensis* is very similar to subspecies *hebridensis* and essentially differs only statistically. A little bigger and more robust, with a larger and looser

flower spike, longer bracts (often over 8mm) and a narrower lip. The lateral sepals are rarely held horizontally, and the spur is stouter. Confined to Cornwall, where it is found on the N coast on stabilised dunes at Lelant golf course and cliffs near St Ives and at Tintagel. **Var. *albiflora*** has unmarked white flowers (rarely washed cream or pink) and unspotted leaves. It is widespread but rather uncommon.
Var. *okellyi* has narrow leaves and flowers which are sometimes a little smaller or broader than typical plants and often almost white or creamy with very faint pink or lilac markings, although in some areas a substantial proportion (often the majority) have darker and heavier markings; the leaves are spotted or unspotted. It is found on the W seaboard of Ireland, the Isle of Man and in N Britain; individual white-flowered plants within these populations are well-nigh identical to var. *albiflora* (and vice versa).

A lot of controversy surrounds *okellyi* with some European authors treating it as a distinct species and other botanists treating it merely as a poorly defined variety. The classic *okellyi* of the literature always has white, almost unmarked flowers and is said to be confined to W Ireland (especially The Burren region of Co. Clare and Co. Galway), the Isle of Man and the W coast of Kintyre in SW Scotland (with single records for Tiree and Sutherland). However, in The Burren and elsewhere these classic white-flowered *okellyi* are just part of a population of plants that are variable in flower colour.

Var. *rhodochila* is an attractive hyperchromic variant with an excess of pigmentation. Lip solidly reddish or purple, usually with a narrow paler pink or white border. The leaves may be more heavily spotted or in extreme cases entirely washed purple. Widespread but rare. **HYBRIDS** Frequent, especially with Southern Marsh and Northern Marsh Orchids. See p.151. **INTER-GENERIC HYBRIDS**
X *Dactylodenia heinzeliana*, the hybrid with Chalk Frangrant Orchid, is uncommon. Found widely in England and Wales.
X *Dactylodenia st-quintinii*, the hybrid with Heath Fragrant Orchid, is rare, but recorded widely in N England and Scotland. The hybrid with Marsh Fragrant Orchid has been recorded in Hants, Lancs and Cumbria; it has no formal name.
X *Dactyloglossum mixtum*, the hybrid with Frog Orchid, is rather rare. Widely scattered records.

HEATH SPOTTED ORCHID

Dactylorhiza maculata

IDENTIFICATION

Often very common, especially in the N and W. Closely related to Common Spotted Orchid and largely replacing it in more acid habitats. Height 4–50cm; has been recorded up to 75cm tall but such robust plants are probably hybrids. A dainty and very attractive orchid, variable in stature but often just 10–25cm high, especially in exposed or relatively dry sites. Flowers very variable but often *very pale* and marked with *fine dots and dashes* but sometimes with *bold lines and loops*. The *broad* lip resembles a voluminous petticoat, with the side-lobes *much larger* than the small, tooth-like central lobe. The leaves usually have numerous fine dark spots. **Similar species** Common Spotted Orchid is rather similar but can be separated by its much more deeply- and evenly-lobed lip and blunter leaves (see also pp.152–155). Hybrids are common, particularly with Northern and Irish Marsh Orchids **Flowering period** Mid May–July, sometimes from the second week of May or rarely to early August. Typically flowers earliest in drier situations and later in wetter habitats.

HABITAT

Heathland and moorland, especially the damper, more peaty areas on the margins of valley bogs, flushes and mires and the raised and slightly drier areas within *Sphagnum* bogs. Also found on acid grassland (for example *rhos* pastures in Wales and culm grassland in SW England), damp unimproved meadows and hill pastures. Usually grows on mildly acidic soils; in areas of chalk or limestone bedrock suitably acidic conditions may be provided by pockets or blankets of peat or areas of drift (superficial deposits of sands and gravels). May sometimes grow in neutral or even slightly alkaline marshes and fens and in such habitats can be very robust. Not as tolerant of shade as Common Spotted and is much less often found in light woodland and even then usually on the edge or along rides. Recorded from sea level up to 915m (Perthshire).

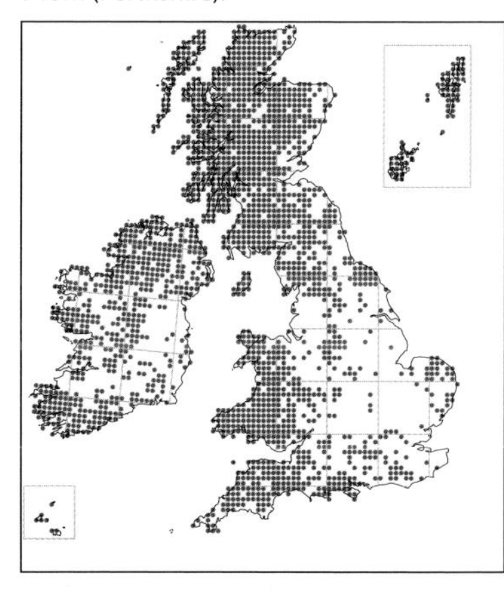

POLLINATION & REPRODUCTION

A variety of insects have been observed carrying pollinia, especially bees and flies. Seed production is good (45% of flowers set seed in one Scottish study).

DEVELOPMENT & GROWTH

Probably similar to Common Spotted Orchid, seedlings appearing above ground less than two years after germination, the first flower spike three years later.

STATUS & CONSERVATION

One of the commonest and most widespread orchids, found throughout Britain and Ireland, but with a bias towards the N and W. There has been some decline, especially in lowland areas in England, and now occupies *c.* 75% of its historic range; many of the losses are relatively recent. Agricultural changes and the destruction of heaths and bogs have caused losses, as has the abandonment of heathland. Tolerates, and even benefits from some grazing but in N Britain suffers, as does much native vegetation, from overgrazing.

DESCRIPTION

See also p.152. **STEM** Pale green, washed purple towards tip and slightly ridged. **LEAVES** 0–1 basal leaves (or sheaths); 2–4 (1–5) sheathing leaves, largest up to 20mm wide (25mm), tips sometimes hooded; variably marked with solid dark spots, fairly frequently unspotted. **BRACT** Green, often washed purple around edges and towards tip, lanceolate, tapering to a fine point, 1.5–2x longer than ovary on lower flowers but around equal in length higher up. **OVARY** Green, cylindrical, 6-ribbed and twisted. **FLOWER** Lateral sepals lanceolate, asymmetrical, marked with darker spots and lines; upper sepal and petals slightly shorter, forming a hood over the column. **LIP** Shorter than wide (*c.* 5–9.5mm long x 6.5–13mm wide), roughly circular or circular–diamond shaped, divided into three lobes, side-lobes much larger than central lobe, often with serrated or frilly edges; central lobe small, triangular, rarely more than 1mm longer than side-lobes. Lip held flat, slightly dished or with side-lobes slightly deflexed. **SPUR** 1/2–1x length of ovary. **SCENT** Faint. **SUBSPECIES** British and Irish plants are conventionally treated as subspecies *ericetorum*; the nominate subspecies, *D. m. maculata*, is found in much of continental Europe. **VARIATION Var. *concolor*** has an excess of pigmentation, with a solidly dark reddish-purple lip, the leaves may also be washed purple. 1–2 records only. **Var. *leucantha*** has pure white flowers. It is widespread but rare. **HYBRIDS** Frequent, especially with Northern Marsh and Irish Marsh Orchids. See p.151. **INTER-GENERIC HYBRIDS** **X *Pseudorhiza bruniana***, the hybrid with Small White Orchid, was recorded from Orkney in 1977. **X *Dactylodenia evansii***, the hybrid with Heath Fragrant Orchid, is found rarely in N England and Scotland. **X *Dactyloglossum conigerum***, the hybrid with Frog Orchid, is very rare; 4 records from N Britain.

EARLY MARSH ORCHID

Dactylorhiza incarnata

IDENTIFICATION

The most widespread marsh orchid. Overall rather scarce, but can be very locally abundant. Height 20–40cm (7–65cm). Probably our most variable orchid, both in flower colour and in stature – from just 5cm tall and very petite to a robust giant of 60cm. The six subspecies display very varied flower colour:

D. i. incarnata Flowers usually whitish-pink. Fens and marshy meadows throughout the British Isles.

D. i. coccinea Flowers deep red. Scattered on both E and W coasts in dune slacks plus some inland sites, especially in Ireland.

D. i. pulchella Flowers purplish-pink or unmarked white. Bogs, especially on acid heaths in S England.

***D. i. cruenta* 'Flecked Marsh Orchid'** Flowers mid–dark pink, frequently also bold spots on leaves and bracts. Fens in W Ireland and NW Scotland.

D. i. ochroleuca Flowers plain creamy. Fens in E Anglia. Very rare.

D. i. gemmana Large, with relatively large, lightly marked flowers. Poorly-known, fens.

Despite the variation in flower colour, relatively distinct. Flowers *small* – almost disproportionately so on robust plants compared to the size of the stem, bracts and ovaries. The flowers appear *narrow* because the sides of the lip are often *folded downwards*, sometimes sharply so, especially as the spike matures. This is further accentuated by the lateral sepals, which are held *vertically* above the flower. Whatever their colour, the flowers have the same basic pattern of markings (with the exception of the unmarked flowers of subspecies *ochroleuca*, subspecies *gemmana* and albinos): a scatter of wriggly lines of varying length in the centre of the lip bounded by a *solid dark line* that forms a *double loop*, with few, if any, dark markings outside the loop.

Similar species The pale pink flowers of subspecies *incarnata* and deep red of *coccinea* are distinctive – no other British marsh orchid shows their coloration, but the various purple-flowered forms can be more confusing. See pp.152–155.

Flowering period Mid May–late June (–late July). Considerable variation both within individual colonies and between sites, but late May and early June is probably the peak for many. Beyond a tendency for plants in the S and W to flower earlier and plants in wetter sites and at higher altitudes to flower later, there seems little difference between the flowering times of the various subspecies.

HABITAT

Subspecies *incarnata*: damp or wet grassland on calcareous or neutral soils. Especially characteristic of unimproved wet meadows on floodplains, but also found in spring-fed fens and flushes, even in the mountains of Scotland. Ironically, although usually found in wet areas, plants cannot survive prolonged submergence. Other habitats include dune slacks (where it may be found with the red-flowered subspecies *coccinea*), old fly-ash tips, occasionally old

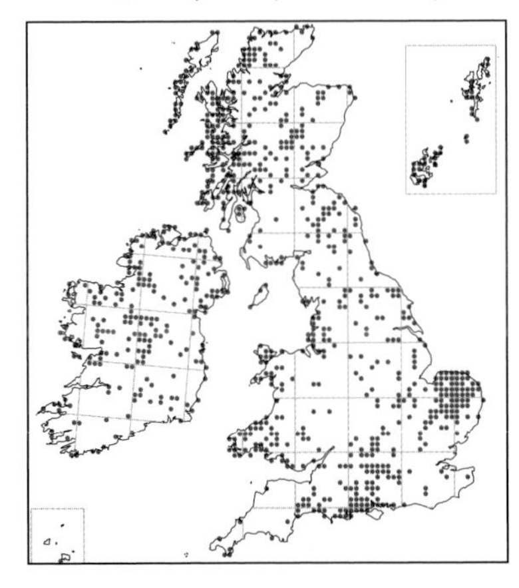

chalk quarries where soil compaction may cause seasonal waterlogging and, very rarely, chalk grassland. Recorded up to 440m above sea level (Perthshire; plants not referred to any subspecies have been found at 610m in Perthshire and Angus).

Subspecies *coccinea*: damp dune slacks, machair grasslands in the Hebrides, slumped clay cliffs in E Norfolk flushed by springs, and damp lake shores in Ireland.

Also recorded from alkaline *Leblanc* waste in Lancs and pulverised fly-ash waste.

Subspecies *pulchella*: valley bogs and acid marshes on heathland, often growing among *Sphagnum*.

Subspecies *cruenta* 'Flecked Marsh Orchid': the limestone region of W Ireland in calcium-rich fens around loughs and turloughs, often growing in the hollows amongst slabs of limestone. In Scotland more-or-less neutral flushes and mires up to 450m above sea level.

Subspecies *ochroleuca*: calcareous spring-fed fens, growing on a mossy carpet in the more sparsely vegetated areas. Suitable sites were created when peat was cut for fuel and the workings subsequently flooded. Such 'turf ponds' slowly fill with vegetation which over time forms a floating mat or 'hover'. Eventually the build-up of material dries them out and they become unsuitable.

POLLINATION & REPRODUCTION

There is no nectar, and pollination is by bumblebees that are duped into visiting the flowers. The process is efficient and seed-set is good.

DEVELOPMENT & GROWTH

Seed probably germinates in the spring, and the first leaves appear a year later. In cultivation plants flower aged 4–5 years. Individual plants have been recorded living for up to 25 years after their first appearance above ground.

STATUS & CONSERVATION

Early Marsh Orchid and its relatives are often known as the 'diploid marsh orchids' because their chromosome count is $2n = 40$ (compared with $2n = 80$ in the 'tetraploid marsh orchids', a group that includes all the other British and Irish marsh orchids). Although of little use in the field, it represents a fundamental division.

Frequently found in small numbers

◀ Subspecies *pulchella*

and often the first of the marsh orchids to vanish if its habitat dries out due to drainage or other changes. Overall, the species has gone from 43.5% of its historical range in Britain and 39% in Ireland. Has declined significantly in lowland areas as riverside meadows have been ploughed, drained or otherwise 'improved', water tables have fallen due to abstraction, or meadows scrub-over. These factors have led to the near-extinction of subspecies *ochroleuca*, which is Nationally Rare and listed as Critically Endangered, with a total of fewer than 60 plants in recent years. Formerly found in the Waveney Valley on the Norfolk–Suffolk border and at Chippenham Fen in Cambridgeshire, it is now extinct at all sites with the exception of Chippenham Fen and one of the Waveney Valley fens in Suffolk, where it was rediscovered in the late 1990s.

The red-flowered dune subspecies *coccinea* has suffered declines due to the scrubbing-over of sand dunes and also coastal development, but still occurs in very large numbers at some favoured sites.

Subspecies *cruenta* 'Flecked Marsh Orchid' is only found at three sites in Scotland and is classified as Endangered in Britain. It is rather commoner in Ireland but with the rapid changes occurring there, including in the Burren region, could be vulnerable to development and 'improvement'.

DESCRIPTION (subspecies *incarnata*)

See also p.153. **STEM** Bright yellowish-green or apple green. **LEAVES** Bright yellowish-green or apple green. 3–5 sheathing leaves, rather broad (the largest more than 2cm wide, sometimes as much as 3.5cm. **BRACT** Apple green, sometimes flushed rose-pink or purplish, lanceolate and fairly long – up to twice as long as ovary and projecting well beyond flowers in lower spike. **OVARY** Green, 6-ribbed and twisted, also bent through *c.* 45°. **FLOWER** Lateral sepals oval to strap-shaped, slightly asymmetric, often with dark pink spots or sometimes ring-shaped markings; upper sepal and petals more oval (petals a little smaller), forming a tight hood. **LIP** Slightly to moderately lobed, the sides usually strongly turned downward, with dots and dashes contained within a more-or-less unbroken dark double loop. **SPUR** *c.* 1/2–1x length of ovary, but less than 7.5mm long. **SUBSPECIES** Genetic analyses show that the subspecies are very similar with the exception of *cruenta*, which does show some genetic differences. The dividing lines between the various subspecies are not hard and fast and intermediates occur. ***D. i. incarnata*** The 'typical' subspecies. In S England the vast majority of plants have pale pink flowers but at a few sites both pale pink and purplish-pink plants can be found growing together. This variation in flower colour appears to be commoner in Wales, Cumbria, Scotland and Ireland;

Subspecies *pulchella* ➤

these 'mixed' colonies are mostly or always in neutral or base-rich habitats.

D. i. coccinea Often relatively small, height 5–20cm (–30cm); can appear stout, squat and apparently stemless. Flowers a distinctive deep red with even darker markings (but less obvious than in other subspecies due to reduced contrast with the dark red ground colour; the markings on the lip are rather narrow and appear quite faint). Bracts rather long, washed and edged purple, upper stem and ovaries also strongly washed purple. Leaves broad-based but sharply tapering, dark green and unspotted. In most dune colonies there are also plants with flowers which are paler and pinker, intermediate with subspecies *incarnata* (and often some typical pale pink *incarnata* as well).

D. i. pulchella Flowers purplish-pink, fading to white around mouth of spur (colour rather similar to Southern Marsh Orchid although slightly more purple). Upper stem, bracts and ovaries variably washed purple; as in subspecies *incarnata* lateral sepals may have ring-shaped marks. Apart from flower colour differs from subspecies *incarnata* in thicker, bolder and more complete dark loop markings on lip, lip less obviously lobed, side lobes either not folded downwards or only folding as the flower gets older – the lip may be flat or even slightly concave. Leaves average narrower and a darker and deeper green. Bracts shorter and less prominent. Some variation in flower colour, and in some colonies very pale yellowish or white flowers are relatively common (vars. *ochrantha* and *leucantha)*.

The name *pulchella* is often used for any purple-flowered Early Marsh Orchid wherever it grows. I prefer to reserve it for the heavily marked, purple-flowered populations growing in acid bogs and to treat the purple Early Marsh Orchids that grow with pale pink-flowered plants in alkaline habitats as colour variants of subspecies *incarnata*.

***D. i. cruenta* 'Flecked Marsh Orchid'** Flowers mid–dark pink, lacking the purple tones of subspecies *pulchella*. 30–65% of plants have dark purplish-brown spots on the leaves, becoming denser towards tip and sometimes merging. If present, spots on the underside are paler, smaller and sparser. Stem, bracts and ovaries more

◀ Subspecies *incarnata*

strongly washed purple than subspecies *pulchella* and sometimes spotted or flecked darker; sometimes ring markings on bracts (almost never spotted in *pulchella*), and lateral sepals more often have ring markings. Lip with broader, bolder loop markings but few dashes within loops, these markings often spread over most of the lip (extending over no more than 2/3 of lip in *pulchella*); lip 4.5–7.5mm long x 4.5–9mm wide, margins often slightly wrinkled, more distinctly 3-lobed, side-lobes moderately reflexed. Stem slender, leaves held stiffly erect, spike looser and less crowded than subspecies *incarnata*, especially in Scottish plants.

Controversy surrounds this subspecies and doubts are sometimes expressed as to whether British and Irish plants are the same as European '*cruenta*'. Strangely, it is sometimes suggested that while Scottish plants are the same, Irish plants are not, being merely spotted-leaved variants of subspecies *pulchella*.

D. i. ochroleuca Often relatively large, sometimes to 70cm tall, with broad stem and large leaves usually held very erect. Bracts large, the lowest usually larger than 30mm x 20mm. Flowers creamy to pale yellow with relatively large lip, *c.* 9mm wide x 7mm long, deeply 3-lobed with *notches on side-lobes*. Has often been confused with pale-flowered variants of other subspecies (e.g. var. *ochrantha)*, but has larger and more distinctly lobed flowers.

D. i. gemmana Recorded from E Norfolk and E Galway with similar plants found elsewhere. Ignored or forgotten for many years but recently resurrected. Robust, to 50cm tall (sometimes 80cm), with six or more leaves; large flowered, lip more than 8.5mm wide x 7mm long, spur usually longer than 7.5mm. Flowers either pink or purple, lip marked with fine dots rather than a loop (suggesting a hybrid origin); may be best to treat it as a large, late-flowering variant of subspecies *incarnata*.

Subspecies *incarnata* ➤

Variation As well as the various subspecies, several varieties have been named. In theory, each different subspecies could be found as each different variety, giving around 24 different combinations:

Var. *punctata* is usually rather small with a few small dots on leaves towards tip. Rare, recorded from Yorkshire, the New Forest and Isle of Coll, among subspecies *incarnata* and *pulchella*. **Var. *leucantha*** has unmarked white flowers; rare. **Var. *ochrantha*** has unmarked pale creamy flowers grading to pale yellow at base of lip (rarely green) – similar to subspecies *ochroleuca* but flowers smaller and less distinctly 3-lobed. Scarce, usually found among subspecies *pulchella*, although also recorded among *incarnata*, *coccinea* and *cruenta*. A hyperchromic variant with an excess of pigmentation and solidly dark lip has also been recorded, albeit rather rarely.

Hybrids Most or all of hybrids with other members of the genus *Dactylorhiza* are sterile and only found as isolated individuals or in small groups. See p.151.

Inter-generic Hybrids The hybrid with Heath Fragrant Orchid has been recorded from four widely scattered localities.

▲ Subspecies *pulchella* var *leucantha*

▲ Subspecies *coccinea*

Subspecies *ochroleuca* ▼

Subspecies *cruenta* ▼

SOUTHERN MARSH ORCHID

Dactylorhiza praetermissa

IDENTIFICATION

Locally common to abundant in England S of a line from the Ribble to the Humber, with a few scattered colonies a little to the N. Also found in S and SW Wales and the Isles of Scilly. Height 20–50cm (–70cm, exceptionally 95cm, but the largest plants probably show some hybrid influence); dwarfed plants, just 10cm high, may occur in marginal habitats such as chalk grassland. Usually fairly distinctive, with purplish-pink flowers that often appear a little washed out or 'dusty', a broad, rounded lip that is only indistinctly lobed – if obvious the central lobe is small and tooth-like – and marked in the centre with fine dots and short dashes. Very variable in stature: some are very short with small, few-flowered spikes, while others are robust with a stout stem, numerous rather broad, unmarked green leaves and a large, many-flowered spike. In addition, two forms of Southern Marsh Orchid are markedly different: Var. *junialis* ('Leopard Marsh Orchid'), has large ring-shaped spots on the leaves and a boldly marked flower. It is hard to distinguish from some hybrids with Common Spotted Orchid. Subspecies *schoenophila* is effectively identical to Pugsley's Marsh Orchid, but their ranges do not overlap (see Variation). **Similar species** See pp.152–155. Hybrids are a major identification headache. Frequently hybridises with Common Spotted Orchid, the hybrids usually tall and robust with the sheathing leaves shorter and blunter than Southern Marsh Orchid and the non-sheathing leaves narrower and more pointed. Importantly, the leaves are usually faintly spotted, and the flowers are intermediate between the two parents. **Flowering period** Late May–early July, occasionally mid May–mid July (even to early August).

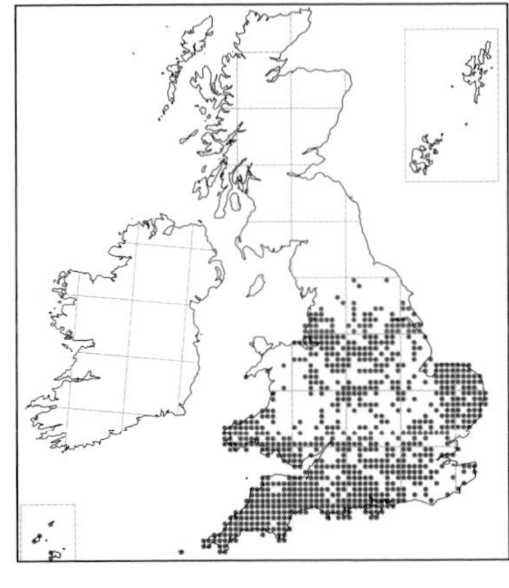

HABITAT

A variety of habitats on alkaline to neutral or even acidic soils, usually (but not always) moist or wet, including damp meadows, fens, marshes, the less acid parts of bogs and wet heathland, dune slacks, marshy gravel pits, road verges and also old industrial sites (especially in NW England and the West Midlands – waste alkali, colliery and fly-ash tips). Also found in old chalk quarries where compacted ground may lead to water-logging. Occasionally grows, often in a dwarf form, on dry chalk grassland. Cannot tolerate being submerged for long periods, and floods lasting more than a month may cause severe declines or even kill all the plants.

POLLINATION & REPRODUCTION

Pollination is efficient and large quantities of seed are produced. Vegetative reproduction, although far less important, may sometimes produce groups of plants.

DEVELOPMENT & GROWTH

The period between germination and flowering is 2–3 years.

STATUS & CONSERVATION

Fairly common, although in the modern agricultural landscape necessarily local. Big colonies stand out but odd plants in marshy meadows may be easy to miss.

Has inevitably suffered quite significant declines as agricultural changes and development, especially the draining and ploughing of flood-plain meadows and pastures, have destroyed its habitats. Has vanished from 20% of the former range, with much of the loss relatively recent.

One of the tetraploid marsh orchids, the chromosome number is 2n = 80.

DESCRIPTION

See also p.153. **STEM** Stout (usually over 5mm in diameter), green. **LEAVES** Mid green, sometimes slightly tinged greyish-green. 4–6 (3–9) sheathing leaves, often slightly more crowded towards base of stem; oblong-lanceolate, more-or-less flat; largest usually 2–3.5cm wide but can be only 1.5cm. 1–3 narrower and more pointed non-sheathing leaves higher on stem. **BRACT** Green, often washed purple, about twice length of ovary and projecting beyond flowers. **OVARY** Green, variably washed purple, cylindrical, 6-ribbed and twisted. **FLOWER** Purplish-pink. Lateral sepals lanceolate, asymmetrical, occasionally blotched darker. Upper sepal and petals lanceolate, petals slightly shorter than sepals, forming a loose hood over column. **LIP** 9–12mm long x 8.5–14mm wide, roughly circular (if slightly 'squashed') and indistinctly 3-lobed; side-lobes broad and rounded, central lobe small; lip purplish-pink, paler towards base and white around mouth of spur, finely marked. **SPUR** 2/3–4/5x length of ovary. **SUBSPECIES** ***D. p. schoenophila*** was described in 2012 to cover all the populations previously placed with Pugsley's Marsh Orchid from E Anglia and S England (see p.179). It is found in the same specialised habitat – alkaline fens low in nutrients – as Pugsley's Marsh Orchid and appears to be more or less identical with it, to the extent that they may only be separable on range. This treatment, based on limited sampling, is not yet definitive, and subspecies *schoenophila* may be returned to Pugsley's Marsh Orchid. **VARIATION Var. *junialis* 'Leopard Marsh Orchid'** has large dark ring-shaped spots (sometimes bars) on the leaves and unbroken dark loops or horseshoe-shaped marks on the lip. Found most often in the S and SE but uncommon. Both Leopard Marsh Orchid and Southern Marsh x Common Spotted hybrids can have hollow, ring-shaped spots on the leaves (although hybrids are more likely to have solid dark spots). The presence of such 'annular' spots is thus not diagnostic. Leopard Marsh Orchid is, however, relatively small and slender, with fewer leaves (averaging smaller and more slender than typical Southern Marsh).

Hybrids are often robust, with many leaves, especially many non-sheathing leaves. Leopard Marsh also has a lip that averages narrower, up to 10mm wide (broader in hybrids, 10mm or more – sometimes much more). Its lip is often more obviously 3-lobed than typical Southern Marsh Orchids, but not as distinctly 3-lobed as hybrids, with dark marking confined to the central zone inside the heavy double loops, and just 1–2 faint markings outside this area – in hybrids dark markings extend over most of the lip. It is possible that 'Leopard Marsh Orchid' results from the leaking of Common Spotted Orchid genes into populations of Southern Marsh, via hybridisation and back-crossing.

Var. *macrantha* has a loose flower spike and large flowers (the lip usually much more than 7.5mm long x 9.5mm wide) with an obvious central lobe that exceeds the lateral lobes by more than 1mm. Rare, found mostly at sites where typical Southern Marsh and 'subspecies *schoenophila*' occur together; it may be an intermediate between the two. **Var. *albiflora*** has white flowers but is very rare.

▲ Var. *junialis* 'Leopard Marsh Orchid'

HYBRIDS Hybrids are frequent, especially with Common Spotted Orchid. See p.151.

INTER-GENERIC HYBRIDS X *Dactylodenia ettlingeriana*, the hybrid with Marsh Fragrant Orchid, has been found in Hants and Glamorgan. **X *Dactylodenia wintoni***, the hybrid with Chalk Fragrant Orchid, has been found in Hants and Surrey.

◄ Subspecies *schoenophila*, Norfolk

NORTHERN MARSH ORCHID

Dactylorhiza purpurella

IDENTIFICATION

The common marsh orchid north of a line from Swansea to Hull (with a very few isolated records from S England). Height 10–20cm (5–45cm). Relatively distinctive. Flowers *deep 'velvety' magenta* with a *distinctive crimson tone* most obvious when freshly opened; the dark markings on the lip are a *lustrous dark crimson*. The lip typically appears *diamond-shaped with straight sides*, especially to the base of the 'diamond' (the lip is actually 3-lobed but the side lobes are folded upwards at the edges to form the diamond). Leaves either unspotted or with a few small spots, except in parts of W Wales, N England and NW Scotland where var. *cambrensis* has dark spots all over the leaves. **Similar species** See pp.152–155. Hybrids with both Heath Spotted and Common Spotted Orchids are relatively common and can be abundant where the two parent species grow together. **Flowering period** Late May–late July (exceptionally from mid May) but mostly early June–mid July.

HABITAT

Damp or wet sites on alkaline, neutral or slightly acid soils, such as marshy fields, roadside verges, lake margins, fens, flushes, seepages along coastal cliffs, dune slacks, machair and sometimes in less acidic peat bogs or in open, damp woodland. An opportunist, it may colonise old quarries and urban 'brownfield' sites, including derelict industrial areas and old waste tips, such as alkaline *Leblanc* waste in Lancs. Sometimes found on drier substrates such as rubble and has been found in gardens. Recorded up to 610m above sea level.

POLLINATION & REPRODUCTION

No specific information, but probably pollinated by bumblebees, as these have been recorded as frequent visitors, especially queens of *Bombus terrestris*. Pollination is efficient and seed-set is good (52% of flowers setting seed in one Scottish study).

DEVELOPMENT & GROWTH

No specific information.

STATUS & CONSERVATION

Quite common and sometimes found in large numbers. There have been losses due to habitat destruction (drainage, ploughing-up of pastures, etc.) but the

overall boundaries of the range remain stable. Has declined by 21% in Britain, with many of the losses comparatively recent, and by 46% in Ireland.

DESCRIPTION

See also p.154. **STEM** Green, washed purple towards tip, ribbed, often stout. **LEAVES** Mid-green to dark green, unspotted or sometimes a few very small spots near tip; 4–8 lanceolate sheathing leaves, largest 1.5–2.5cm wide; 1–2 non-sheathing leaves higher on stem (sometimes none, rarely up to four) and a small basal leaf. **BRACT** Green, washed purple (and sometimes spotted on ribs), cylindrical, 6-ribbed and twisted. **OVARY** Relatively large and inflated, even when flowers fresh, greenish-yellow, prominently ribbed and slightly twisted. Upright but narrows at top into a stalk-like base for the flower which is bent through more than 90°, holding flower pendant. **FLOWER** Deep magenta, whiter around mouth of spur. Sepals roughly oval, marked with irregular dark reddish-purple rings and lines. Petals a little shorter and unmarked. Lateral sepals held at *c.* 45°; upper sepal and petals form a loose hood over the column. **LIP** 5–8mm long x 6–10mm wide, variable in shape, but usually shallowly 3-lobed with a small central lobe. Heavily marked dark crimson. **SUBSPECIES** None. **VARIATION** **Var. *cambrensis*** has leaves boldly spotted dull purple; spots are distributed over the whole surface of the leaf and can be sparse or occasionally so numerous that they coalesce into a solid dark patch; spotting is absent in a small minority of plants. Bracts spotted or washed purple. The flowers average slightly paler and pinker than normal and therefore the dark markings are more contrasting. Lip a little larger, 6–8.5mm long x 9–11mm wide, rather distinctly 3-lobed, often with notches between the lobes, with the side-lobes slightly reflexed. Tends to be taller than typical plants with longer and narrower leaves but with the spike smaller in proportion to the whole plant. Grows near to the sea in W Scotland, including the Outer Hebrides and Orkney. In W Wales found rather rarely in dune slacks and flood plain meadows on the coast from Cardigan to Anglesey. Also recorded from SE Yorkshire and recently from several sites in Cumbria. (It is possible that Var. *cambrensis* represents the leaking of Common or Heath Spotted Orchid genes into populations of Northern Marsh, via hybridisation and back-crossing.) **Var. *albiflora*** has white flowers. **Var. *atrata*** is a hyperchromic variant with an excess of pigmentation. Lip solidly dark magenta, lacking spots or lines but sometimes with

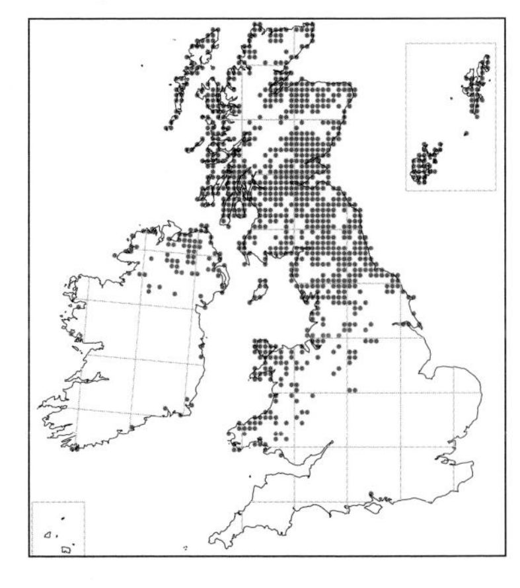

The typical form ➤

well-defined paler margin. Leaves heavily spotted (rarely an overall purple wash), with a few small spots on their lower surfaces. Known only from damp fields at Hartlepool which may be contaminated with heavy metals. **Var. *crassifolia*** has broad, fleshy leaves and a large lip. Rare. **Var. *maculosa*** has many small dots all over the leaves and sometimes a paler lip. Rare, recorded from SE Scotland. **Var. *pulchella*** has the lip marked with spots and dashes rather than bold loops. Recorded from Scotland. **HYBRIDS** Frequent, especially with Common and Heath Spotted Orchids. See p.151. **INTER-GENERIC HYBRIDS** **X *Dactylodenia varia***, the hybrid with Heath Fragrant Orchid, has been found in N England, Scotland and Co. Down. The hybrid with Marsh Fragrant Orchid has been found once in Cumbria (no formal name). **X *Dactyloglossum viridellum,*** the hybrid with Frog Orchid, is very rare; recorded from Co. Durham, Scotland and Ireland.

➤ Hybrid Northern Marsh x Common Spotted Orchid *Dactylorhiza* x *venusta*, showing the coloration and short, sack-like spur of a marsh orchid, and the 3-lobed lip and bold lip markings suggestive of Common Spotted Orchid.

PUGSLEY'S MARSH ORCHID

Dactylorhiza traunsteinerioides Other names: **Narrow-leaved Marsh Orchid**

IDENTIFICATION

Scarce and very local – an enigmatic species almost always found in species-rich wet alkaline fens. Height 10–30cm (6–40cm; the most robust in Ireland). A slender, delicate marsh orchid. Flowers variable in colour but often purplish-pink. Hard to identify, but a suite of features give it a subtle but distinctive appearance.

* Flower spike usually *distinctly one-sided*, all flowers facing roughly the same way.
* Relatively *few, well-spaced flowers*, usually only 6–14 (2–18) per spike.
* Relatively *large flowers*. The lip is at least 7.5mm wide and usually wider, up to a maximum of *c.* 13mm (the average of several plants should be at least 9mm).
* Lip usually obviously 3-lobed, with the central lobe longer and rather narrower than the rounded side-lobes and usually with prominent notches between the lobes; the side-lobes turned downwards and the central lobe *projects as a prominent 'tooth', itself also sometimes turned downwards* (flatten the lip from below with a finger in order to measure the width and accurately judge the shape). In a natural position the lip often has a characteristic triangular or deltoid shape, narrower at the base and broadest at the tip; a shape accentuated by the side-lobes being more strongly reflexed at the base.
* Lip well-marked with dark dots, loops and squiggles, the markings *often extending right to the edge*.
* *Few leaves*. The total of 3–5 includes 0–1 bract-like non-sheathing leaves; do not count the short basal leaf hidden at the bottom of the stem.
* Leaves relatively narrow, with the second leaf from the bottom (usually the broadest) 6–15mm across (rarely to 18mm and usually averaging no more than 12mm).

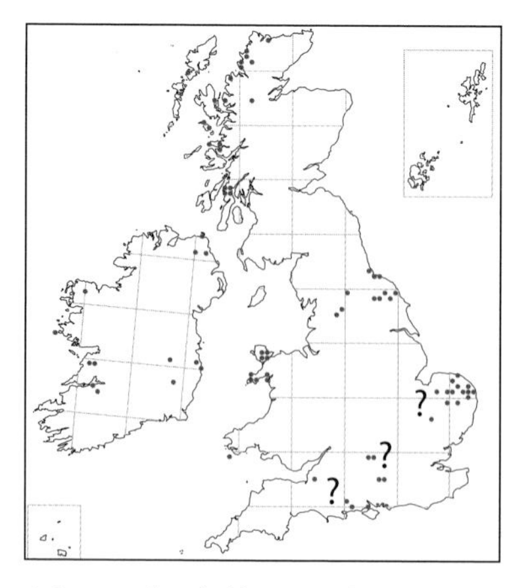

* Leaves often held erect and more-or-less evenly spaced, rather than being gathered into a rosette.
* Leaves mostly unmarked but at a few sites in Yorkshire and Ireland the plants have a scatter of fine dark spots, and in Scotland subspecies *francis-drucei* ('Lapland Marsh Orchid') has very boldly marked leaves and bracts.
* Upper stem and especially bracts washed purple.

Populations of marsh orchids in Norfolk, N Suffolk and scattered sites elsewhere in S England (N Somerset, N Hants, Berks and Cambs), long placed with Pugsley's Marsh Orchid, were described in 2012 as subspecies *schoenophila* of Southern Marsh Orchid. The switch of these populations from one species to another was based on genetic research, a fast developing field, and may not be definitive. Importantly, they are found in the same alkaline fen habitat as Pugsley's Marsh Orchids elsewhere, and look pretty much identical too. If the move to Southern Marsh Orchid is accepted, Pugsley's may only be identifiable on range, a most unsatisfactory situation.

In order to confirm the identification it is best to examine a selection of plants to get a feel for the range of variation. It is unlikely that a Pugsley's Marsh Orchid would occur as a single individual among a colony of commoner species and there should always be a few candidates to look at. **SIMILAR SPECIES** See pp.152–155. **FLOWERING PERIOD** Late May–late June (exceptionally from early May). Flowering can begin a little later, towards early June in the N, with 'Lapland Marsh Orchid' sometimes in flower until July.

HABITAT

Very specific. Wet fens and flushes where the ground water is 'base-rich' due to the influence of chalk or limestone. Grows in the wettest areas among a relatively open community of sedges, rushes and scattered reeds, rooting into the mossy layer at the base of the taller vegetation. Almost always associated with Black Bog-rush. Some sites are fairly extensive but in others the correct conditions are restricted to small patches within a larger area of marshland. In Ireland it has been recorded rarely from dune slacks. Only exceptionally found in very slightly acid conditions and in Scotland 'Lapland Marsh Orchid' may spread from its base-rich flushes into nearby wet heathland, which is slightly more acid. Found up to 370m (NW Yorkshire), with subspecies *francis-drucei* recorded up to 310m (Kintyre).

POLLINATION & REPRODUCTION

Poorly known. Probably pollinated by bees – although other insects such as flies may remove the pollinia, they do not go on to pollinate other orchids successfully. In studies of Pugsley's and 'Lapland' Marsh Orchids in W Scotland, only 17–35% of flowers set seed, reflecting low levels of pollination, probably due to a lack of suitable pollinators and poor weather. Each ripe capsule contained *c.* 3,000 viable seeds. Vegetative reproduction may also be possible.

DEVELOPMENT & GROWTH

No specific information.

STATUS & CONSERVATION

Nationally Scarce. The strongholds are in N Yorkshire, Anglesey and the Llyn Peninsula in Caernarvonshire. Scarce and local in Ireland, and just three sites in Northern Ireland. Rare in Scotland, where both the typical form and 'Lapland Marsh Orchid' have a broadly similar distribution in the west, including the Inner and Outer Hebrides. At many sites, especially in Scotland, there may only be small numbers of flower spikes. Two good clues for locating the species are flowering period and habitat. Pugsley's Marsh Orchid is one of the first marsh orchids to come into flower, and it is almost always found with Black Bog-rush (a rather large, tussock-forming member of the sedge family).

A tetraploid marsh orchid, with the chromosome number 2n = 80. All of the tetraploid marsh orchids are thought to have originated from ancient hybridisation events involving the ancestors of Early Marsh Orchid and Common or Heath Spotted Orchids. Pugsley's Marsh Orchid is intriguing in that the same parent species have crossed on several occasions, in the British Isles, the Alps, Scandinavia and perhaps elsewhere, and each hybridisation event has given rise to a similar plant. After much debate, genetic research suggests that, although very similar, each population has a distinct ancestry and represents a separate lineage and they are thus best treated as separate species.

British and Irish plants are therefore an endemic species, *D. traunsteinerioides*.

In the lowlands, Pugsley's Marsh Orchid has disappeared due to the destruction and drainage of its habitat. Even where wetlands remain, often protected as SSSIs and reserves, a general lowering of the water table (due to drainage and abstraction) can cause a degradation of the habitat. In the uplands, sheep or deer graze most sites. This can keep the vegetation open and prevent a succession to scrub but it limits the number of flowers produced, sometimes severely so. When sites are not grazed the number of flowering plants has increased, although this may not reflect the number of seeds that are able successfully to germinate. More direct threats include forestry and drainage, and only a few of the 30 or so populations of 'Lapland Marsh Orchid' are protected in any way at all.

Pugsley's Marsh Orchid has undergone a moderate decline in Britain but losses have been greater in Ireland, at 67.5% of the total historical range, and its numbers appear to have fallen there more than almost any other orchid.

DESCRIPTION

See also p.154. **STEM** Green, usually washed purple towards flower spike. Slender, averages 2.3–3.5mm in diameter (rarely more than 5mm). **LEAVES** Unspotted or occasionally with a few, well-scattered, faint, purplish-brown spots or transverse bars *c.* 1mm across; uppermost non-sheathing leaf may be washed purple. One short basal leaf and 2–4 (–5) sheathing leaves, longest rarely more than 12cm long (17cm in robust Irish populations); broadest leaf usually 6-15mm wide (rarely to 18mm); moderately keeled, sometimes slightly hooded at tip. **BRACT** Lanceolate, long (lower bracts longer than flowers), green, variably but often strongly washed reddish-purple. **OVARY** Green, strongly washed purple; 6-ribbed. **FLOWER** Sepals oval-lanceolate, lateral sepals asymmetrical, very occasionally with darker markings; petals slightly smaller and more oval; upper sepal and petals form a hood over the column. **LIP** Relatively large, 6.5–9.5mm long x 7–13mm wide, usually a flattened oval divided into three obvious lobes, side lobes variably folded downwards. Variable shades of purplish-pink, becoming whitish at mouth of spur, with an irregular

◀ Lapland Marsh Orchid

'Lapland Marsh Orchid' ➤

pattern of dark lines, dots and blotches. No correlation between the ground colour of the lip and the extent and colour of the markings; dark plants in Yorkshire may have relatively small dot and dash markings whereas paler flowers on Anglesey may have long, heavy and contrastingly dark lines and blotches. **Subspecies *D. t. francis-drucei* 'Lapland Marsh Orchid'** Bold, dark purplish-brown spots, blotches, bars and rings on upperside of leaves and bracts, which may also be edged purple (and bracts often spotted below). 0–2 non-sheathing leaves which may also have a few small marks on underside. Lip with intense dark purple or crimson lines, rings and spots, sometimes merging to form dark patch in centre; lateral sepals marked with dark rings, elongated spots and dots. Averages slightly shorter than the typical form (6–18cm tall, sometimes to 24cm), but Scottish populations of typical plants are also small. (WCA Schedule 8.) Very local in W Scotland, including the Inner and Outer Hebrides; has also been identified in Co. Antrim in Northern Ireland. Sometimes found with typical Pugsley's Marsh Orchids. **Variation** Rather variable. The size and shape of the leaves, presence or absence of leaf spotting, number of flowers and their colour and markings all vary, both between colonies and within them. **Var. *albiflora*** has white flowers. Very rare. On the basis of genetic data, **Hebridean Marsh Orchid** has been placed with Pugsley's Marsh Orchid as subspecies ***francis-drucei* var. *ebudensis***. If further data prove this to be correct, it would be fascinating, as Hebridean Marsh Orchid occurs in a very different habitat and looks rather different too. Pending confirmation, I prefer to continue to treat it separately (pp.184–186). **Hybrids** See p.151. **Intergeneric Hybrids** The hybrid with Heath Fragrant Orchid has been recorded once, in Cumbria.

A typical plant from Yorkshire ➤

HEBRIDEAN MARSH ORCHID

Dactylorhiza ebudensis

IDENTIFICATION

Known only from the N coast of North Uist in the Outer Hebrides, where locally abundant in a 'metapopulation' extending SW for *c.* 4 km from Port nan Long, with a couple of very small satellite populations a few km to the W (also a few plants on Berneray). Height 4.5–20cm. Usually rather short and squat with a distinct *purple* tone to the flowers. The leaves vary from *more or less plain to solidly blackish-purple*, but most have *dark rings or fine blotches* (plants with darker leaves tend to have a darker ground colour to the flower). **Similar species** Only likely to be confused with Northern Marsh Orchid, but distinguished as follows:

- Flower spike looser, less symmetrical and often one-sided (symmetrical and compact in Northern Marsh).
- Lip relatively larger, more spreading, and more obviously 3-lobed; the side lobes seldom bend upwards at the edges in the manner that gives Northern Marsh its characteristic diamond shape.
- Flowers usually a clearer, purer purple (when freshly opened the flowers of Northern Marsh have a distinct crimson tone; on older flowers the colour can become more purple, but the dark lip markings retain a deep crimson lustre).
- Leaves often (but not always) more heavily marked, with ring-spots and blotches, also finely edged with purple (especially the uppermost), even when not spotted. Northern Marsh may have fine, regular spotting, but never extensive dark markings. Var. *cambrensis* of Northern Marsh, which has heavily spotted leaves, has been recorded elsewhere in the Outer Hebrides, but its spots tend to be smaller, more even in size and more evenly distributed than in Hebridean Marsh.
- Whole plant averages smaller and more petite.

Flowering period Late May to late June.

HABITAT

Found on the machair, a unique, species-rich, coastal grassland habitat that develops on low calcium-rich dunes in W Scotland and Ireland.

POLLINATION & REPRODUCTION

No specific information.

DEVELOPMENT & GROWTH

No information.

The taxonomy of Hebridean Marsh Orchid is confusing. Until recently it was grouped together with Irish Marsh Orchid and var. *cambrensis* of Northern Marsh Orchid into one species named 'Western Marsh Orchid *D. majalis*'. Genetic studies then suggested that Hebridean and Irish Marsh Orchids should be treated as distinct species and that Hebridean Marsh Orchid originated as a cross between Early Marsh Orchid, subspecies *coccinea*, and Common Spotted Orchid, subspecies *hebridensis*. More recently, again on the basis of genetic data, Hebridean Marsh has been placed with Pugsley's Marsh Orchid as var. *ebudensis* of 'Lapland Marsh Orchid' (i.e. of subspecies *francis-drucei*). It has a rather different ecology, however, and looks somewhat different too, and intriguingly an enclave of Pugsley's Marsh grows in its typical habitat surrounded by Hebridean Marsh Orchids. I suspect this might not be the end of the story and pending a definitive answer I prefer to maintain its separate identity.

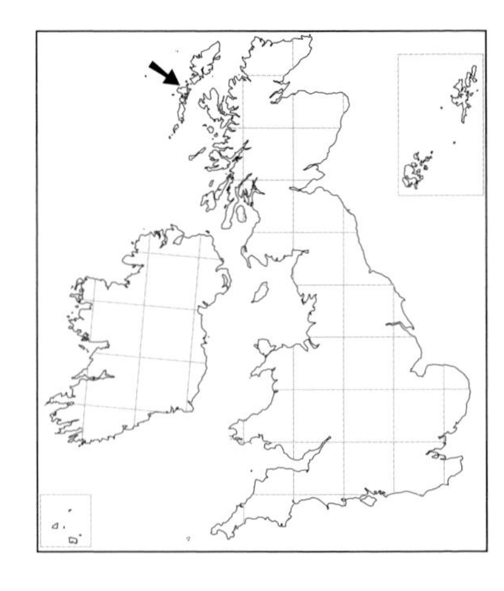

STATUS & CONSERVATION

Nationally Rare and listed as 'Vulnerable'. Much of its range lies within the Machairs Rabach and Newton SSSI and there seems to be no immediate threat to the species, but such a limited distribution means that it will inevitably always be vulnerable. One of the tetraploid marsh orchids, with a chromosome number of 2n = 80.

DESCRIPTION

See also p.155. **STEM** Green, heavily washed with dark purple towards tip. **LEAVES** 2–3 (–4) sheathing leaves, 10–14mm wide, and a single lanceolate non-sheathing leaf. Variably, often heavily marked with brownish-purple rings or fine blotches. **BRACT** Lanceolate, lower slightly longer than flowers but becoming shorter towards tip of spike; green, heavily washed purple or sometimes entirely purple and often spotted darker. **OVARY** Ribbed, twisted and very heavily washed purple. **FLOWER**. Sepals and petals elliptical, with lateral sepals slightly asymmetric; upper sepal and petals (which are shorter) form a hood over the column. **LIP** Broader than long, 6–8.5mm long x 8–12mm wide, and distinctly 3-lobed with central lobe longer than side-lobes; rich purple-magenta, whiter towards throat of spur, heavily marked darker. **SUBSPECIES** None. **VARIATION** A hyperchromic variant has occasionally been found; an excess of pigmentation produces a solidly dark lip with a paler border. **HYBRIDS** See p.151.

IRISH MARSH ORCHID *Dactylorhiza kerryensis*

IDENTIFICATION

Endemic to Ireland, where scattered but widespread and locally common. Height 10–30cm (–40cm). Rather variable in stature – dwarf plants of coastal grasslands can look rather different to the robust populations in damper and more sheltered spots. A compact spike of purplish-pink, well-marked flowers identifies this species as one of the marsh orchids. The lip has three *rounded* lobes and is *usually heavily marked, often with double loops*, while the leaves may be *either spotted or unspotted.* **SIMILAR SPECIES** In Ireland three other marsh orchids need to be eliminated: Early, Northern and Pugsley's. See pp.152–155. Early Purple Orchid flowers at the same time as Irish Marsh Orchid, sometimes in the same grassy habitats, and also has heavily spotted leaves. Its leaves are arranged in a floppy rosette at the base of the stem, however, and its flowers are much less heavily marked, with just a few dark spots in the centre of the lip. **FLOWERING PERIOD** Mid May–mid June, sometimes into July.

HABITAT

Varied but mostly on neutral or slightly alkaline soils and including wet meadows and pastures, road verges, lough shores, the edges of acid bogs, dune slacks, damp hollows on short, closely grazed grassland and dry grassy slopes near the sea.

POLLINATION & REPRODUCTION

No specific information.

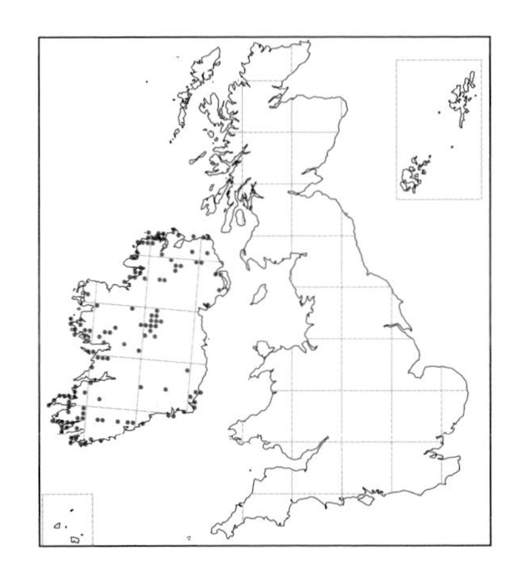

DEVELOPMENT & GROWTH

No specific information.

STATUS & CONSERVATION

Generally fairly common and sometimes found in large numbers. Although mapped for Northern Ireland, its occurrence there is disputed. Vulnerable to agricultural changes, especially with European Union money funding agricultural 'improvement' in Ireland, but so far losses appear to be modest. The difficulties of marsh orchid identification and the relatively low level of botanical recording in Ireland may be obscuring any changes in its status.

Until recently part of a conglomerate of marsh orchids that took the name 'Western Marsh Orchid *Dactylorhiza majalis*'. The conglomerate included *D. majalis* from mainland Europe as well as Irish Marsh Orchid, Hebridean Marsh Orchid and the *cambrensis* subspecies of Northern Marsh Orchid. Recent research, including the analysis of DNA, has suggested that Irish and Hebridean Marsh Orchids are unique endemic species and that *cambrensis* from W Wales belongs with Northern Marsh Orchid. *D. majalis* of Europe does not occur anywhere in the British Isles.

One of the tetraploid marsh orchids, the chromosome number is 2n = 80.

DESCRIPTION

See also p.155. **STEM** Green, variably washed purple, ridged towards tip. **LEAVES** Sheathing leaves usually 15–25mm wide (12–30mm). In *c.* 50% of plants leaves spotted brownish-purple. Rarely, lower surface of leaf spotted, or it has a purple rim. **BRACT** Fairly long, green, variably washed purple (sometimes strongly so) but only occasionally spotted. **OVARY** Slender, twisted and ridged, green, often heavily washed purple. **FLOWER** Lateral sepals oval, asymmetrical, variably marked darker with dots, squiggles, or rings (sometimes unmarked). Upper sepal oval, petals slightly shorter, narrower and more pointed; together they form

a hood over the column. **LIP** Pinkish-purple (slightly deeper, more intense and closer to purple than Southern Marsh Orchid), becoming paler towards centre and white around mouth of spur. Usually wider than long, 7–11mm long x 9–14mm wide, with three distinct rounded lobes; usually heavily marked, often with double loops. **SUBSPECIES** None. **VARIATION** Plants with unspotted leaves and bracts ('**var. *kerryensis***') are found in W and SW Ireland, sometimes in 'pure' colonies, but are scarce. Their flowers are, on average, relatively paler and less intense and the lip is often flat and marked with dots and dashes that do not form a pattern of loops. They also tend to be slightly shorter and stockier and are said to flower a little later, into July. This form is sometimes treated as a distinct species, '*D. kerryensis*', due to its floral characters and later flowering, and allied with Southern Marsh Orchid, which it resembles in flower colour and pattern although the dark spots on the lip are more numerous and more extensive. **Var. *occidentalis*** is the more widespread variety, with relatively intensely-marked leaves and flowers. A hyperchromic variant with an excess of pigmentation and a solidly dark lip has been recorded rarely. **HYBRIDS** See p.151.

EARLY PURPLE ORCHID *Orchis mascula*

IDENTIFICATION

Height 10–45cm (–60cm). One of the most widespread orchids, although rather local in many areas. The combination of early flowering, purple flowers and spotted leaves is distinctive. Plants with unspotted leaves are not uncommon, however, especially in Scotland, and plants with white or rose-pink flowers occur occasionally. The structure of the flower should prevent any confusion. The lateral sepals are held erect as 'angel's wings' and the spur is relatively long and curves gently *upwards*. **Similar species** Green-winged Orchid is easily separated by the parallel green stripes on its sepals. In addition, the 'hood' of the flower is formed by all the sepals and petals and it always has unspotted leaves. Marsh orchids have flowers of various shades of purple but have a short and often bag-like spur that is either straight or curves downwards, and often heavier and more extensive dark markings. If present, their leaf spots tend to be regular and often elongated sideways rather than irregular and often elongated lengthwise as in Early Purple.
Flowering period Early April–early June in the S, exceptionally from mid March, but mostly late April–late May. Averages a little later in upland areas and Scotland, occasionally continuing until early July.

HABITAT

Very variable. Grows on a variety of soils, but has a preference for chalk, limestone or boulder clay and avoids acid conditions. Found in a wide variety of old grasslands, both dry chalk downland and damp hill pastures, as well as meadows, rocky mountain ledges, railway embankments and cuttings, road verges, grass-covered dry-stone walls and limestone pavements. Also grows in deciduous woodland, usually the better-lit areas along rides, tracks and edges. Particularly associated with coppice woodland, where often a big increase in the number of flowering plants the second or third year after coppicing but a decline thereafter as the canopy closes again. Does not colonise new sites easily. Usually found in ancient woodland rather than plantations or secondary woodland, unless woodland and scrub have invaded old orchid-rich grassland; conversely, many colonies on road verges and banks may be relicts of long-gone woods. Recorded up to 880m above sea level (Angus).

POLLINATION & REPRODUCTION

Pollinated by bumblebees and to a lesser extent cuckoo bees and a variety of solitary bees. A visiting insect touches the column and the pollinia are stuck to its head by the sticky viscidia. Once the bee has left the flower, the pollinia on the insect rotate forward in *c.* 30 seconds to be in position to make contact with the stigma of the next flower visited. Self-compatible, and sometimes self-pollinated.

The flowers lack nectar and offer no reward to a pollinator. Rather, it is thought that the orchid takes advantage of the naivety of the bumblebee queens which, newly emerged from hibernation in the spring, have yet to learn which flowers are

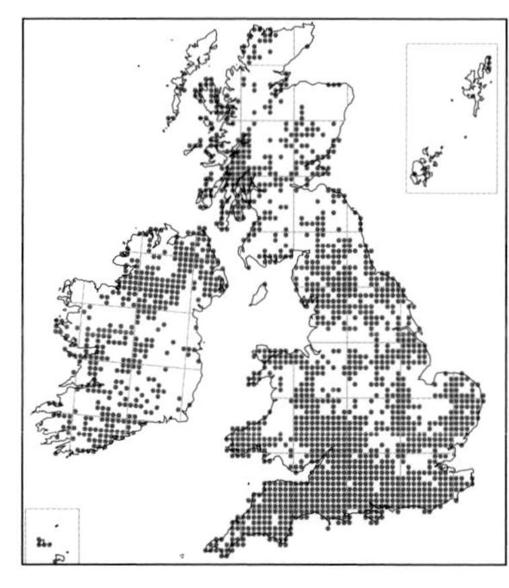

genuine sources of nectar. These naive bees are attracted, at least for a while, by the bright colour and scent of the flowers. Various species of bee emerge from hibernation at different times, and the orchid can thus take advantage of a succession of pollinators. Pollination rates are highest when the orchids grow amongst other, genuinely nectar-bearing plants, which act as 'magnet species'. Seed-set is variable; when poor, the lowest, earliest-opening flowers are most likely to be pollinated. Vegetative reproduction may occur occasionally via the production of additional tubers.

DEVELOPMENT & GROWTH

The first aerial leaves are usually produced in the fourth year after germination, and more and bigger leaves appear in successive seasons until enough reserves have been accumulated to produce a flower spike; up to eight years may elapse between first appearance above ground and first flowers. Around 60% of plants will flower in successive years; the remainder either appear as a rosette of leaves or, in *c.* 17% of cases, spend a year dormant underground before appearing again; up to 12 years dormancy has been noted in Europe, and individual plants have been recorded flowering over at least nine years.

STATUS & CONSERVATION

Fairly common or even abundant in some areas (e.g. in the White Peak of Derbyshire, Yorkshire Dales and the Burren in W Ireland), but may be very local on acid soils and largely absent from some regions, such as the Fens, S Lancashire and parts of W Wales, and very scattered in others – the Borders, NE Scotland, Orkney, Shetland, the Outer Hebrides and SE Ireland.

The third most widespread orchid in the British Isles (after Common Spotted and Heath Spotted Orchids). Has vanished, however, from 28% of its historical range in Britain and 21% in Ireland, and the decline appears to be ongoing in Britain. Losses are due to the destruction or 'coniferisation' of woodland and, perhaps more importantly in recent years, the loss of permanent grasslands as pastures and meadows have been ploughed and reseeded. In most of lowland Britain, away from reserves, has been lost from farmland and is now largely confined to ancient woodland and to marginal sites, such as road verges and churchyards, that have escaped agricultural improvement. In the N and W overgrazing may be a problem as tolerant of light grazing only.

DESCRIPTION

Stem Stout, pale green, angled and usually flushed purple towards tip. **Leaves** 3–8 basal leaves held close to the ground, either spreading upwards and outwards or in a flatter rosette; variably oblong-lanceolate and often blunt-tipped, glossy green, usually marked with large, irregular, rounded or elongated blackish-

purple spots on upper face (rarely also on underside). 2–3 rather smaller and more pointed sheathing leaves higher on stem that may have a few spots or a purple wash. **Spike** Oval or cylindrical but often rather irregular in shape and rather open, especially in lower half, with 10–50 flowers, occasionally more. **Bract** Green, usually strongly washed purple, lanceolate, about as long as the ovary, which they clasp. **Ovary** Green, variably washed purple and clearly ribbed. **Flower** Various shades of purple, occasionally pale rose-pink or white. Sepals lanceolate, lateral sepals asymmetrical, petals smaller and more arrow-shaped; upper sepal and petals form a hood over the column, the two lateral sepals held erect and pushed backwards (may almost touch at rear). **Lip** Points downwards and outwards, 3-lobed, side-lobes folded downwards (entire lip sometimes folded along its centre line); central lobe usually the largest, shallowly notched; edges of all three lobes crinkled. Base of lip and mouth of spur much paler and whiter, sometimes with some yellow tones, and usually spotted purple, the spots composed of dense tufts of short papillae. Spur long and narrow (at least as long as ovary), broadening a little towards blunt tip and curving upwards. **Column** Variably greenish or purple, pollinia dark green. **Scent** Initially sweet-smelling, recalling honey or Lily-of-the-valley, but quickly becomes rank, usually said to recall a tomcat's urine and especially pungent at night. **Subspecies** None. **Variation** **Var. *alba*** lacks anthocyanin pigments and has unmarked white flowers with yellow pollinia, green stem and bracts, and no leaf spots. It is rather scarce. An even scarcer variant has white flowers with purple spots at base of lip and sometimes spotted leaves, while a very rare 'broken-coloured' variant has a pale pink lip copiously flecked with tiny purplish marks. **Hybrids** ***O.* x *wilmsii***, the hybrid with Lady Orchid, has been reported very rarely in Kent. **Inter-generic Hybrids X *Anacamptorchis morioides***, the hybrid with Green-winged Orchid, has been recorded rarely and sporadically.

LADY ORCHID *Orchis purpurea*

IDENTIFICATION

Height 20–50cm (–100cm). Confined to Kent (with a few outliers in Sussex, Hants and Oxon), where very locally common. Identification straightforward. Usually rather large and statuesque, with dark reddish-purple flower buds. One of the 'manikin' orchids, the flowers form a miniature human figure: the sepals and petals form a dark 'bonnet' that, together with the unopened buds, contrasts strongly with the whitish lip, which is divided into several lobes to form the arms and the 'skirt' of the lady. **Similar species** Burnt Orchid is superficially similar, with dark buds, dark hood and a white, purple-spotted lip, but it is very much smaller, seldom more than 15cm tall. **Flowering period** Early May–early June, exceptionally from mid–late April, but generally at its best mid–late May.

HABITAT

Woodland, both ancient woodland and secondary woods, almost always on thin, well-drained chalky soils (rarely on limestone or other calcareous substrates). Favours beechwoods and often grows on south-facing slopes, frequently on banks or on the 'terraces' formed by the root plates of trees, either among a carpet of Dog's Mercury or on bare leaf-litter. However, its preferred habitat may actually be scrub or coppice, and it does not flower so freely in shade, being happier in open, well-lit situations – along paths and rides, in clearings and on the lower edges of woods. Indeed, may cease to flower and 'disappear' if shade becomes too dense, only to spectacularly reappear after coppicing, tree falls or felling opens up the canopy. Conversely, although often found just outside a wood, rarely occurs in full sun on open downland. Also likes shelter from wind, and in exposed situations the leaves and flowers may be scorched in a cold spring, the leaves often turning yellow.

POLLINATION & REPRODUCTION

Pollinated by small flies and bees, including small digger wasps. Seed-set is variable, sometimes very low, with only 3–10% of flowers producing ripe capsules. However, in some years and at some sites it can be good. Most reproduction is by seed although vegetative reproduction also takes places, forming clumps of plants.

DEVELOPMENT & GROWTH

From germination to first flowering takes 8–10 years and the pattern of development is typical of the genus *Orchis*. Plants may live for at least another ten years, flowering at least three times in that period but seldom every year; the remains of the previous year's dried, dead spike is not often seen next to the current flowers. There are usually large numbers of non-flowering plants in any population.

STATUS & CONSERVATION

Nationally Scarce and listed as Endangered. Locally frequent on the North Downs in Kent, with over 100 sites in two areas: the downs either side of the Medway Valley and sporadically eastwards towards the

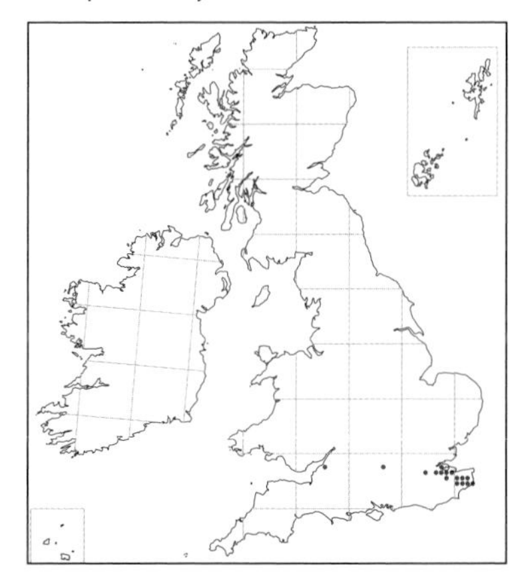

has long been noted that Lady Orchids are subject to the depredations of rabbits and deer, which nip off the flowers and attack the leaves, as well as slugs. Deer are now commoner in England than at any time for a thousand years, making this an increasingly serious threat.

DESCRIPTION

Stem Green, becoming purplish-brown towards tip. **Leaves** Green, shiny or almost 'greasy', 3–5 (–7), broad, oval to oval-oblong; lower blunter and form a basal rosette, upper successively more pointed, keeled and clasping. Leaves appear above ground mid January–mid February. **Spike** Oval to oblong, and lax or densely flowered. Robust plants have up to 50 flowers. **Bract** Greenish to purple, tiny, elongated and scale-like. **Ovary** Bright green, sometimes washed purple along the six ribs, distinctly twisted. **Flower** Sepals oval, petals shorter, much narrower and

Stour Valley; the downs between the E slope of the Stour Valley and Dover. Elsewhere, two sites in Oxon (including Hartslock, where *c.* 25 plants of non-British origin, but it is not known whether they were deliberately planted/sown), a site in Hants and a single plant in East Sussex.

Lady Orchid has been lost from 57% of the historic range. It was formerly found in Surrey, mostly prior to 1930, West Sussex, where at least five sites up to1976. Recorded from Herefordshire in 1967, and Leigh Woods in the Avon Gorge, Somerset, in 1990 (although this plant quite possibly originated from Lady Orchids in cultivation at the University of Bristol Botanic Gardens a short distance away).

Past declines were due to the loss of woodland and the cessation of coppicing. The population in Kent is now stable and some colonies are large (over 3,000 plants were counted at one site in recent years) although others are declining, and it does not easily colonise new areas. It

more strap-shaped but broaden to a spear-shaped tip; together they form a hood; pale green, irregularly blotched on both surfaces dark purple or purplish-brown, blotches becoming more numerous and coalescing towards tip, base and sides. **Lip** Pale pink to white, washed violet or rose around edges, variably spotted pink to reddish-purple (spots formed by tufts of tiny papillae). Deeply lobed with two long, narrow arms and two broad terminal lobes, often with frilled edges, forming the 'skirt' (the latter usually have a tiny tooth between them). Spur cylindrical, 1/4–1/2x length of ovary, curved and pale green blotched purple. **Column** Pale green washed pinkish or purplish; pollinia blotched purple. **Scent** Variably reported to smell of vanilla or bitter almonds or to be unscented. **Subspecies** None. **Variation** Relatively variable. Plants growing in the open tend to be shorter and darker-flowered than those in woods, with a tendency towards brown rather than purple markings. There is also much variation in the shape, colour and markings of the lip; may be white and even unspotted in some plants and heavily washed pink with dark purple spots in others. The hood may also vary from paler to darker shades of purple.

Plants in W Kent differ slightly from those in E Kent (from the eastern side of the Stour Valley to Dover), averaging shorter (20–38cm tall rather than 30–76cm) with a shorter, denser flower spike and shorter ovaries (13–19mm rather than 19–25mm); lips more heavily spotted, washed rose to purple (rather than salmon to brownish-red), blunter and less deeply lobed. These two groups of populations have not, however, been given names.

Var. ***albida*** lacks anthocyanin pigments and has a pure white lip and a distinctive white or straw-coloured hood with green veins. It is scarce. **Var.** ***pseudomilitaris*** has narrower and more reddish lobes on the lip, resembling the arms and legs of a Military Orchid. The hood is, however, the normal reddish-purple. It is rare.

Hybrids *O.* x *wilmsii*, the hybrid with Early Purple Orchid, has been reported very rarely in Kent. ***O.* x *meilsheimeri***, the hybrid with Man Orchid, was found in E Kent in 1998. ***O.* x *angusticruris***, the hybrid with Monkey Orchid, appeared in Oxon in 2006 (see p.204). Hybrids with Military Orchid have also been reported.

MILITARY ORCHID *Orchis militaris*

IDENTIFICATION

Very rare. Confined to two sites in the Chilterns (Homefield Wood in Bucks and a site in Oxon), and Mildenhall in Suffolk. Height 20–40cm (5–60cm). One of the so-called 'manikin' orchids in which the flower resembles a tiny human figure. The sepals and petals form a soldier's helmet, purple-striped on the inside, the lip has four lobes, two for the arms and two for the legs, and the purple spots down the centre of the lip are reminiscent of buttons. The allusion to the military was coined before soldiers habitually wore red uniforms and may refer to the resemblance of the hood to an ancient 'coal-scuttle' helmet. **Similar species** Monkey Orchid is similar in the general structure of the flower, but its legs are kinked, narrower and do not broaden towards the tip, and the hood formed by the sepals and petals is more open. Also, its flower spike is not only shorter and more crowded but also more jumbled and disarrayed, and all the flowers open at roughly the same time. **Flowering period** Mid May–mid June; at its best in late May and early June. Once a flower has been pollinated it usually shrivels within a day.

HABITAT

Grassland, scrub, woodland glades, woodland edges and, formerly, rough fields, always on chalk. Does best in light scrub on old pastures and in the shelter of woodland edges. Favours some shade and needs bare ground for seedling establishment (rather than a closed sward).

POLLINATION & REPRODUCTION

Probably pollinated by bumblebees and perhaps also hoverflies. Produces no nectar and hence 'cheats' visiting insects. A pollination rate of 3–11% was given for plants in Suffolk but at Homefield Wood 40% and 24% of flowers were naturally pollinated in 1999 and 2000 respectively. In addition, up to 20% of plants at Homefield Wood were hand-pollinated from 1986–98 and hand-pollination can result in almost 100% seed set; hand-pollinated plants have flowered every year for a decade and are thus not 'weakened' by the process. Each capsule contains *c.* 6,000 seeds.

Vegetative reproduction is probably important for the British populations, maintaining numbers when recruitment from seed is low.

DEVELOPMENT & GROWTH

Relatively long-lived. Many will live for ten years after their first appearance above ground and a significant proportion live for at least 17 years. In a study in Suffolk the 'half-life' varied from 2.2–7.8 years (see p.251). In Suffolk the tip of the shoot appears above ground late December–early January and the leaves start to unfurl by early March. Plants are vulnerable to late-winter frosts and mid spring drought.

The period between germination and the appearance of the first aerial shoot is 3–5 years and 30–50% of plants flower in their first year above ground. Once they have appeared, most plants flower at least once and many do so every year, but intervals of up to 11 years between

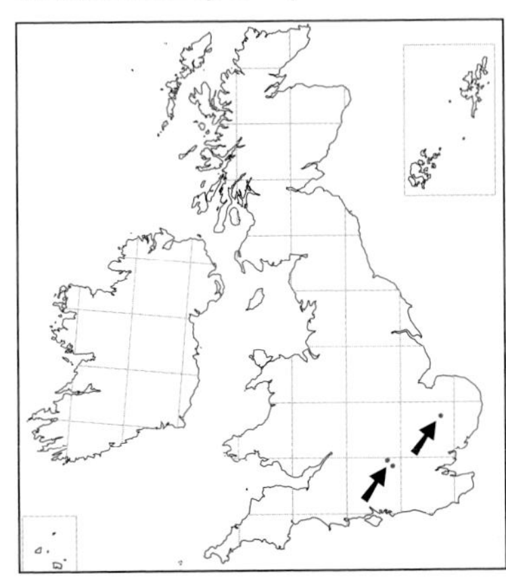

flowering have been noted. Perhaps a third of plants retreat underground during their life, usually spending only one year dormant but in some cases up to three years, and absences of up to eight years have been recorded in Suffolk. In any given year, 5–15% of the adult population can be underground. 'Dormancy' is not a good option for the plants, however, as underground plants have the highest probability of dying.

STATUS & CONSERVATION

Nationally Rare and listed as Vulnerable; WCA Schedule 8. Once fairly common in the Chilterns, but declined dramatically from *c.* 1850 and thought to be extinct after 1929. Several factors contributed to the decline, including the ploughing-up of downland and the collection of specimens, but the main cause may have been a big increase in the number of rabbits.

Military Orchid became something of a Holy Grail for British botanists, and in May 1947 was rediscovered by J. E. Lousley at Homefield Wood in Bucks. There were 18 flower spikes, and Lousley thought that the increase in available light after trees were felled during World War II had probably prompted their appearance. Lousley kept his discovery secret, fearing that the colony would be wiped out by collectors, but in 1956, after a search of all likely sites, it was found by Richard Fitter and Frances Rose. They famously dispatched a postcard to Lousley with the cryptic message 'The soldiers are at home in their fields'. Homefield Wood came to be managed by BBOWT in 1969, but it was not until the end of the 1980s that the location was made truly public. Numbers steadily increased, aided by hand-pollination, to *c.* 80 in 1995, with 45 in flower, then from 1995 there was a dramatic upturn, with over 200 plants, 130 flowering, in 2003, and 180 flowering in 2012.

Military Orchid was found at a second site in the Oxfordshire Chilterns in 1970. The number of flowering plants did not exceed five until 1999 but there was then a rapid increase, with 25 flowering in 2003 (with a further 25 non-flowering) and 324 by 2012. This population was also hand-pollinated for at least 10 years from 1988

As part of the conservation programme, plants propagated at Kew were planted out at Homefield Wood and the nearby Warburg reserve in 1996 (both 1- and 2-year-old tubers). Survival was poor but those that remained first flowered in 2000 and 2002 respectively. In addition, 25 wild plants were transplanted from Homefield Wood in 2000 to a site around 25km away. Survival of these mature plants has been better than young seedlings, with some flowering and setting seed.

In 1954 the Military Orchid was found at Mildenhall in Suffolk, a region from which there were no previous records. The colony was in an old chalk pit (now the Rex Graham Reserve) within a Forestry Commission plantation and in 1954 the surrounding pines were only 1.5m tall. There were at least 500 plants. By 1958 the number had risen to 2,854 and remained at this level until the late 1960s, although only *c.* 10% flowered. However, the population then declined rapidly to 252 plants in 1971, with perhaps only 100 in the following years. To protect the orchids the Forestry Commission had erected a tall fence around the pit in the 1960s. This excluded both deer and people and within a few years Sycamore and Wild Privet had taken over. These were cleared in autumn 1972 and the colony increased slightly and stabilised at 300-400 plants, with about 100 flowering. Scrub clearance continued and the overshadowing pines were finally removed in 1985-86. From 1987 there was a dramatic increase: in 1990 there were 279 flowering spikes and 1,115 plants and in 2000 748 flowering plants and too many non-flowering plants to count. Only 436 flowered in 2007, however, and in an interesting development numbers in the original pit have continued to fall, while the orchid has colonised a purpose-made scrape constructed next to it.

Genetic fingerprinting has shown that the three English colonies are distinct and may represent independent colonisations from Europe.

DESCRIPTION

STEM Green, variably tinged purple towards tip. **LEAVES** Bright shiny green, prominently keeled, strap-shaped or, on more robust plants, broader and more oval, and slightly hooded at tip. 2–5 leaves form a basal rosette held *c.* 45° above the horizontal, with 2–3 sheathing leaves higher on stem; upper part of stem bare. **SPIKE** Oval or conical, becoming cylindrical as the flowers open, with 10–40 (2–57) flowers. **BRACT** Triangular to oval, very short (much shorter than ovary) and green, strongly washed purple or rose. **OVARY** Green, washed purple, boldly ribbed and strongly twisted. **FLOWER** Sepals oval with pointed tips, petals narrower and more strap-shaped; they form a hood, with the sepal tips swept upwards and the hood very open. Sepals pale dove-grey, outer surfaces lightly washed lilac, becoming more purplish at the base (unopened buds pale pinkish-grey), inner surface with bold, longitudinal purplish lines and an irregular purplish wash. Petals more uniformly pale purplish. **LIP** Two narrow, strap-shaped lobes form arms and two shorter and broader lobes form legs, with an additional small pointed projection between the legs; whitish, flushed pink, with arms and legs more-or-less solidly purple. Two rows of purple spots run down centre of lip, formed by tiny papillae, with solid dark purple lines along centre of arms. Spur purple, short (*c.* 1/2 length of ovary), cylindrical, slightly down-curved. **SCENT** A faint vanilla-like scent is sometimes reported. **SUBSPECIES** None. **VARIATION** Suffolk plants may average taller and paler-flowered than Chiltern populations; the latter have been separated as var. *tenuifrons*, but British plants all fall within the normal range of variation found in mainland Europe. **HYBRIDS** ***O.* x *beyrichii***, the hybrid with Monkey Orchid, occurred in Oxon until the middle of the 19th century.

MONKEY ORCHID *Orchis simia*

IDENTIFICATION

Very rare. Confined to Hartslock (Oxon) and two sites in Kent (one near Faversham and Park Gate Down, where introduced). Height 10–30cm (–45cm). Distinctive, the flowers resemble a Spider Monkey: the hood forms the 'head' and the *slender curved lobes* of the lip arms and legs. In most orchids the flowers open from the bottom of the spike upwards, but in Monkey Orchid they often open rapidly from the top downwards, although they may open synchronously (or even from the base upwards in Kentish plants). The *rapidity* with which all the flowers open is characteristic of Monkey Orchid. **SIMILAR SPECIES** In Military Orchid the legs are straight, distinctly broader and widen towards the tip, and the hood forms a longer, neater 'helmet'. The flower spike is taller and less crowded, the flowers opening in sequence from the bottom of the spike upwards. **FLOWERING PERIOD** Mid May–early June; flowering peaks around a week earlier in the Chilterns than in Kent.

HABITAT

South-facing slopes on open, grazed chalk grassland. Probably favours the interface between grassland and woodland or scrub, benefiting from the shelter that scattered trees and shrubs provide from drying winds and grazing animals, and the slightly moister conditions. Should the shade become too dense, however, ceases to flower or even to appear above ground.

POLLINATION & REPRODUCTION

Some natural pollination does occur in England, with the flowers visited by flies, bees and butterflies. Although the spur is not thought to contain nectar, two swellings near its mouth may contain sugars which can be extracted by insects. Seed-set has been poor, however, and some populations have been hand-pollinated. Pollination rates appear to improve significantly once numbers are above a certain threshold – the spectacle of lots of flowers may be more attractive to potential pollinators. In a study in Holland, vegetative reproduction was rare.

DEVELOPMENT & GROWTH

Much of the information on Monkey Orchid comes from a small population in the Netherlands that may not be entirely typical. The first aerial leaf appears 3–4 years after germination and flowers are produced after a further 3–6 years, by which time the plants have at least four basal leaves (a similar timescale has been recorded for Kentish Monkeys). Young plants occasionally disappear underground after they have produced their first leaves and then reappear again after 1–2 years.

Plants can be long-lived, flowering for up to 19 consecutive years, although this may be exceptional. They often 'rest' between bouts of flowering as vegetative rosettes and may even be dormant underground for 1–2 years. If absent for three years, however, they are almost certainly dead. Severe winters inhibit flowering and may result in many plants dying, as the tubers typically lie in shallow soil. May be badly affected by drought and if the leaves are

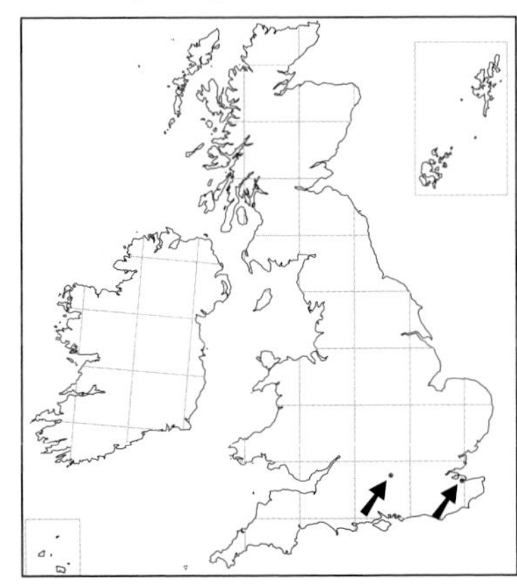

grazed off the plant will not flower the following season – presumably they are unable to build up sufficient reserves.

STATUS & CONSERVATION

Nationally Rare and listed as Vulnerable: WCA Schedule 8. In the 18th and early 19th centuries frequent in S Oxon between Wallingford, Reading and Henley. Declined dramatically, however, from *c.* 1840 due to the ploughing of downland and the collection of specimens, although the main cause of its demise may have been a big increase in the number of rabbits. By the mid 1920s there was only one substantial colony remaining in the Chilterns, at Hartslock, but such was the secrecy surrounding this site that the species was generally thought to be extinct in Britain. In 1949–50, the field that held the Monkey Orchids was ploughed. Fortunately, the upper part of the slope escaped (this is where the orchids now grow); there may already have been a few growing in this refuge, and some tubers rescued from the lower part of the field were replanted there with unknown results. Following this catastrophe, just one Monkey Orchid flowered at Hartslock in 1950–52 and numbers remained painfully low for many years – it was not until 1968 that the population reached even the modest total of eight spikes. In 1975 BBOWT bought Hartslock but, as on most downland sites, scrub encroachment was a problem, leading to a programme of scrub removal, and the area is now grazed in the autumn and winter by sheep. Despite management the population increased very slowly (*c.* 60 plants in the late 1980s, with *c.*1/3 flowering each year), but from1994 numbers started to expand rapidly and by 2010 comprised *c.* 450 plants, with 1/3–2/3 flowering each year. Three factors may have helped: a run of mild winters, which probably reduced mortality and encouraged plants to flower. Hand pollination, which began in 1977 (now stopped, as rates of natural pollination and seed production are high). In 1992 the main colony was fenced against rabbits – it is thought that rabbits had been eating most of the first orchids to emerge (leaves may appear as early as November but more usually January–February).

Despite an apparently healthy population, studies have shown that the genetic variability at Hartslock is very low, as the population had gone through a genetic 'bottleneck' when it was reduced to a tiny handful of individuals. A more recent concern is hybridisation with Lady Orchid, which appeared there in 1999. The hybrids ('Lonkey Orchids') first flowered in 2006 and steadily increased (309 in 2010, with 77 flowering), with fears they may out-compete their parents. In the last few decades Monkey Orchid has appeared sporadically in very small numbers at other sites in the Chilterns.

In Kent, Monkey Orchid was first recorded in 1777 near Faversham and then again in the early 1800s but was not seen again until 1920–23 when a few plants flowered near Canterbury. At the other end of the county, in W Kent, from 1952 onwards a single Monkey Orchid appeared on a disused tennis court at a vicarage at Otford. However, on the retirement of the botanically minded vicar the orchids were moved to nearby private land, where the largest flowered once only, in 1957, and then the Monkeys vanished.

In 1955 a Monkey Orchid appeared near Faversham in Kent (at the current native site), but the single plant was eaten. More were found in the following years and from 1958 until at least 1985 they were hand-pollinated. They steadily increased, with 205 flowering plants in 1965. Then the 1975–76 drought badly affected the colony, with none flowering, but in 1977 plants reappeared and by the mid 1980s there were again 30–50 plants, with *c.* 10 flowering. Numbers are now stable at 200 plus, but only a small proportion flower.

From 1958 onwards seed from Faversham was scattered elsewhere in Kent, and a population became established at Park Gate Down. The first three flowered in 1965, seven years after seed was sown, but not again until 1976. Numbers stabilised at *c.* 100 plants in the mid 1990s but has increased significantly since then.

A few plants appeared at Spurn Point in SE Yorkshire in 1974, over 250km from the nearest known source of seed. Numbers peaked at 25 plants, with nine flowering, but they only persisted until 1983, after which the site was washed away in a storm.

DESCRIPTION

STEM Green, usually washed brownish-purple towards tip, angled, with 2–3 sheaths at base. **LEAVES** 3–4 shiny green basal leaves, oval-oblong, often keeled and blunt-tipped; 2-3 sheathing leaves higher on stem. **SPIKE** Roughly globular and crowded; most have 10–20 flowers but up to 30 (42) on well-developed plants. **BRACT** Very small, *c.* 1/3 length of ovary, triangular, chaffy and whitish. **OVARY** Green, heavily washed purple, boldly 6-ribbed, twisted and curved. **FLOWER** Sepals lanceolate, petals slightly shorter, much narrower and more strap-shaped; together they form a hood that encloses the column, the sepals slightly splayed at the tips. Sepals whitish on outer surface, variably washed pink with irregular violet-purple dots, blotches and streaks. Inner surfaces more heavily blotched and streaked, sometimes almost solidly so. Petals similar but sometimes more solidly washed purplish-pink. **LIP** Four lobes form arms and legs, with a small projection between the legs; the tips of the lobes curve forwards and upwards. Lip whitish in centre, variably flushed violet-purple and spotted violet (spots formed by tufts of papillae), becoming violet-purple towards extremities. Spur pale pink, *c.* 1/2–3/4x length of ovary, slightly down-curved and blunt-tipped. **COLUMN** Reddish. **SCENT** Faintly vanilla-scented. **SUBSPECIES** None. **VARIATION** Kentish plants average taller than those in the Chilterns, with stouter stems, more and bigger leaves and darker flowers. There is much overlap, however. **HYBRIDS** ***O.* x *beyrichii***, the hybrid with Military Orchid, occurred in the Thames Valley until the mid 19th century. ***O.* x *bergonii***, the hybrid with Man Orchid, was recorded in 1985 at Faversham in Kent and in Hampshire in 2013. ***O.* x *angusticruris***, the hybrid with Lady Orchid, appeared in 2006 at Hartslock.

MAN ORCHID *Orchis anthropophora*

IDENTIFICATION

The bulk of the population grows on the North Downs in Surrey and Kent. Rare and very local elsewhere, confined to the region S and E of a line from Bristol to the Humber. Height 20–30cm (15–45cm, rarely to 65cm). The long, narrow spike and very man-like flowers are distinctive. The sepals and petals form a 'hood' and the tiny figure faces downwards, concealing its 'face'. The lip is deeply lobed to form the arms and legs. Flowers yellowish to greenish, variably washed red; plants in full sun may average the reddest. **Similar species** Frog Orchid may be similarly coloured but is usually rather smaller, and the lip is not divided into arms and legs. Common Twayblade is also vaguely similar but its tiny green flowers are rather different in shape and it has only two large, rounded leaves. **Flowering period** Early May–late June (sometimes from late April), but usually at its best in late May. The flowers open in slow progression from the bottom of the spike, so flowering can be protracted.

HABITAT

Typically well-drained grassland on chalk or limestone, often on or at the foot of a slope, with a predilection for abandoned quarries and pits. Roadside verges, churchyards, field margins and stabilised dunes or shingle can also provide suitable habitat. Frequently grows in relatively long, rank grass and among scrub and will sometimes spread under the eaves of nearby woodland. Vulnerable to competition, however, and dense scrub will crowd it out; conversely, populations can be eliminated by heavy grazing.

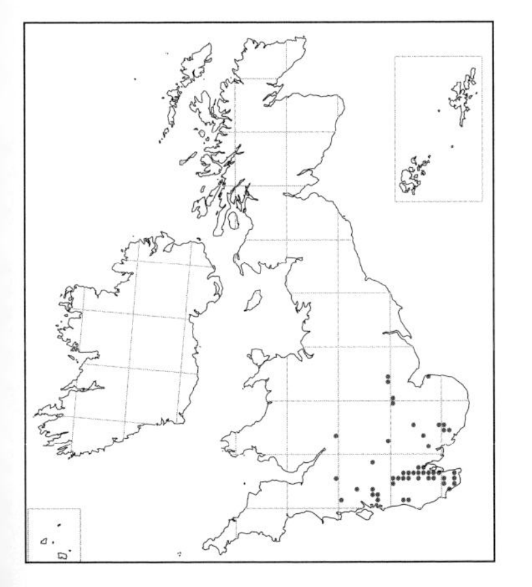

POLLINATION & REPRODUCTION

Little is known about the pollination of this species. However, the numerous hybrids with Lady, Monkey and Military Orchids found in Europe would suggest that it shares a suite of pollinating insects with those species; in England ants and hoverflies have been seen with pollinia on their heads. Seed-set is moderate to good, but nevertheless it is thought that most reproduction is vegetative via the production of additional tubers.

DEVELOPMENT & GROWTH

Once they have appeared above ground for the first time, plants may live for up to 14 years, although they may not flower every year or even appear above ground. Conversely, some may flower for five years in a row. Only rarely dies after flowering just once, and in a study in Bedfordshire the 'half-life' averaged 5.8 years and varied from 4.0–7.8 years (see p.251). There is no information on the period between germination and flowering.

STATUS & CONSERVATION

Nationally Scarce and listed as Endangered. Very localised, and only likely to be found on reserves or protected road verges. Away from the North Downs, 4–5 sites each in Hants and Suffolk, and 2–4 in Essex. Two each remain in Sussex, Northants and Lincs, and just single sites in Wilts, Glos, Oxon, Beds and Norfolk; it was apparently introduced to its site in Warks, in 1968.

Still decreasing in numbers, Man Orchid has been lost from 56% of its historical range. It was formerly much commoner in East Anglia but vanished as pastures were ploughed from the late 19th century on. Sites were also destroyed when quarries and pits were used as landfill, field margins were sprayed or subject to spray drift and road verges cut or sprayed unsympathetically. Sites have also been lost to scrub encroachment.

Formerly placed in the genus *Aceras* as *Aceras anthropophorum*. Indeed, it was the only species in that genus, which was distinguished from the genus *Orchis* by the lack of a spur. DNA studies have confirmed that this difference is purely superficial and that Man Orchid is clearly an *Orchis*.

DESCRIPTION

STEM Pale green with some membranous sheaths at extreme base. **LEAVES** Green, dull or slightly bluish, distinctly veined, keeled and narrowly oval-lanceolate to strap-shaped. A basal rosette of 3–4 leaves, some lying flat and some held at *c.* 45°; higher on stem 1–2 smaller and more lanceolate sheathing leaves. In some areas leaves appear in spring, in others in Nov–Dec, becoming fully formed by Jan–Feb. In all plants the tips of the lower leaves are often scorched by May and they all die off after flowering; this pattern of growth is probably an adaptation to the mild wet winters and hot dry summers in the core area of its distribution around the Mediterranean. **SPIKE** Tall, narrow, more-or-less cylindrical and dense, with up to 50 (90) flowers. **BRACT** Green, lanceolate,

0.5x length of the ovary. **Ovary** Pale green, long, cylindrical, boldly ribbed and twisted. **Flower** Green or yellow, variably tinged red. Sepals oval, various shades of yellowish-green, often with a distinct maroon fringe and midrib. Petals pale green, slightly shorter, much narrower and more strap-shaped. Both sepals and petals form a hood over the column, with the petals fully concealed. **Lip** Variably green or yellow, often strongly washed red or reddish-brown, especially around edges, but can be pure red or yellow. Hangs almost vertically downwards and has three lobes: two long, narrow side-lobes at the base (the arms) and a terminal lobe that is itself divided half way to the base into two lobes (the legs); sometimes a tiny projecting tooth between the legs. No spur, rather two shiny, whitish swellings either side of base of lip which curve round to join the column and enclose a shallow

pit with two small, nectar-secreting depressions. **Scent** A faint, unpleasant smell. **Subspecies** None. **Variation** **Var. *flavescens*** lacks red pigments (anthocyanins) and has a green hood and contrasting yellow lip. It is rare. **Hybrids** In Europe, hybrids with Monkey, Military and Lady Orchid are common but in England such hybrids have only been found twice. ***O. x bergonii***, the hybrid with Monkey Orchid, was found in Kent in 1985–89 (it has been suggested that this may be the result of inadvertent hand pollination), and in Hants in 2013. ***O. x meilsheimeri***, the hybrid with Lady Orchid, was found in Kent in 1998, when two plants were seen.

BURNT ORCHID *Neotinea ustulata*

IDENTIFICATION

Rare, but may be very locally abundant, although colonies can be very localised, with hundreds of plants in one area and few or none elsewhere, even on seemingly suitable ground. Height 2.5–30cm. In S England usually *c.* 4–10cm, with the very shortest plants found on heavily grazed sites. In N England a little taller, usually *c.* 9–12cm, even to 15–25cm when growing among taller vegetation. One of the group of orchids in which the flower resembles a tiny human figure, with the lip divided into arms and short, stumpy legs. The combination of small size, dark reddish-purple buds (giving the top of the spike a scorched or burnt appearance, hence the name) and a white lip marked with fine reddish-purple spots, is distinctive. **Similar species** Lady Orchid is similar in general flower structure and colour but is much larger and its lip differs in the details of shape and coloration. **Flowering period** There are two varieties that differ in their flowering period. The early-flowering form flowers mid May–mid June in S England and may, on average, be just a few days later in the N; it is usually at its best in the last ten days of May. The late flowering form flowers late June–early August.

HABITAT

Ancient short grassland on chalk and limestone, often on south- or west-facing slopes, although the late-flowering populations are not so fussy about the aspect. In S England favours the narrow 'terracettes' that follow the contours of the slope on the chalk downs and also Bronze and Iron Age earthworks where the ground has been undisturbed for centuries. Only on rare occasions will it colonise new sites. For example, at Martin Down in Hants it appeared in the 1980s on grassland that had been under the plough until 1957. Burnt Orchid is also sometimes found in alluvial water meadows where the silt is derived from chalk, e.g. in Lincs and formerly at sites in Hants and Oxon. This habitat was always uncommon in S England but in the past Burnt Orchid was found relatively frequently in riverside pastures in the N.

POLLINATION & REPRODUCTION

Poorly known. Early-flowering plants are pollinated by flies that feed on nectar and sugary plant juices. Presumably the flies are initially attracted by the combination of colour and scent, but there is no nectar to reward them and they must be satisfied in some other way. Once on the flower, the flies work from the uppermost, unopened buds downwards and insert their proboscis in a head-down position. They 'taste' with their feet beforehand and there may be some sugary secretion from the flower to guide them into the 'correct' position; the groove at the base of the lip may in addition act as a 'leading line'. The mouth of the spur is narrow and rather like a keyhole; this shape and the design of the column may be related to the unconventional 'upside-down' position of the pollinating fly. Butterflies also visit the flowers but have not been recorded

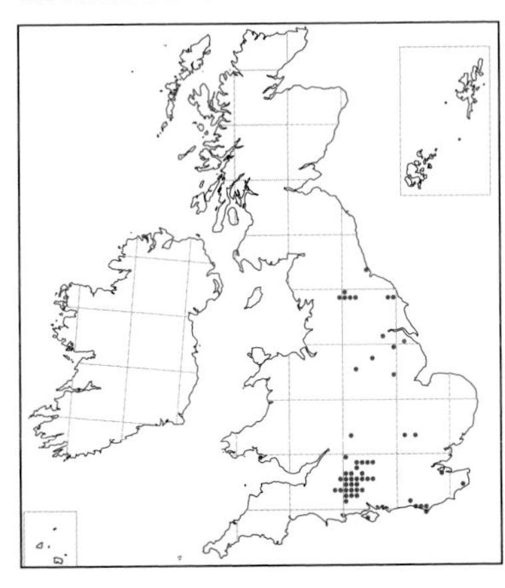

carrying pollinia. With its different scent and later flowering, var. *aestivalis* may attract a different suite of species and, in Europe, beetles have been recorded as pollinators. Whatever its mechanism, pollination is not very efficient in England and seed-set is relatively poor, with *c.* 20% of flowers producing ripe capsules. Nevertheless, most new plants are recruited to the population from seed.

Vegetative reproduction takes place but is thought to be relatively unimportant. Groups of 4–6 spikes are not uncommon, however, and we have seen a compact cluster of 12 (photo below); such groups may well be the product of vegetative reproduction.

DEVELOPMENT & GROWTH

The degree of fungal infection is unclear, with both high and low rates being reported. Dormancy is common, however, and Burnt Orchids are able to spend up to three years (sometimes even four) underground without producing aerial parts. During this period the fungal partner must play a major role (although dormant plants have a significantly reduced chance of surviving). Most plants flower for 1–4 seasons in succession (rarely for up to seven successive years) and then either die or retreat into a period of dormancy underground or as non-flowering rosettes.

Seed germinates to produce a protocorm which grows to a length of 20–30mm before the first root is produced (the longest protocorm known among orchids with a similar pattern of development). Both in cultivation and in the wild Burnt Orchids can flower within about three years of germination.

STATUS & CONSERVATION

Nationally Scarce and listed as Endangered. Once found throughout the chalk and limestone areas of England, the species has undergone a major decline – probably the greatest of any orchid in the last 50 years or so – and is now extinct in many areas. Out of a total historical range covering 265 10km squares, the *New Atlas* records a post-1987 presence in just 55, representing a 79% decline. Most losses can be directly attributed to agricultural changes, from the ploughing of grasslands from the Enclosures of the late 18th and early 19th centuries onwards, while in the latter half

of the 20th century the 'improvement' of grasslands with artificial fertilisers and pesticides continued to take a toll. The remaining grassland sites are vulnerable to scrub invasion due to a lack of grazing, especially in S England, where first rabbits were decimated by myxomatosis and latterly livestock has become less economically viable. Paradoxically, overgrazing is also an issue, especially in N England, where losses due to building and other development have also occurred.

Fortunately many sites are now protected as SSSIs, nature reserves or are on Ministry of Defence land and managed for the benefit of their chalk flora and fauna. For Burnt Orchid the ideal regime is light spring grazing until late April with the return of sheep in late July–late September (when the new leaves start to appear); cattle are unsuitable as they can trample and destroy the plants and also damage the turf.

In S England, of *c.* 350 historical localities only *c.* 75 survived into the 1990s. Most have fewer than 50 flowering plants and at some just 1-2 spikes make sporadic appearances; fewer than ten regularly hold more than 200 plants. The strongholds are now in Wilts, Hants and East Sussex, and notable concentrations include Parsonage Down in Wilts where there may be 30,000 flower spikes spread over 95ha in one continuous 'mega-colony', probably the largest surviving population in NW Europe. Outlying relict populations cling on in Dorset, the Isle of Wight, West Sussex, E Glos, Berks, Herts, Beds and Kent.

Declines have been even more marked in the Midlands and N England, and the species is now only found in Derbyshire, Lincs, N Yorkshire and at a single site on the coast of Co. Durham. In 1993, bucking the trend, Burnt Orchid was recorded for the first time in Wales, on limestone grassland in Glamorgan.

The late-flowering variety *aestivalis* (see Variation) is only found in Wilts (four sites), Hants (five sites) and East Sussex (14 sites, mostly between Lewes and Eastbourne, one holding nearly 1,000 spikes in 2002).

Until recently Burnt Orchid was placed in the genus *Orchis*, but genetic studies have resulted in its transfer to the genus *Neotinea*. There it joins the superficially rather different Dense-flowered Orchid.

◀ Var. *aestivalis* Var. *ustulata* ▶

◀ Var. *aestivalis*

DESCRIPTION

UNDERGROUND The aerial stem grows from a pair of tubers. **STEM** Yellowish-green, slender and ridged towards tip, with 2-3 membranous white sheaths at the very base. **LEAVES** Green with a faint blue tone, fading to pale green by flowering time. A rosette of 2–5 elliptical-oblong keeled leaves at base of stem and 1–2 bract-like leaves higher up. Leaves appear in autumn, are wintergreen and start to wither as the plant comes into flower (by which time non-flowering plants have already vanished). **SPIKE** Dense, initially conical but lengthens and becomes cylindrical when all 15–50 flowers have opened.
BRACT Reddish-purple, lanceolate and rather short, c. 2/3 length of ovary.
OVARY Green, twisted and obscurely 6-ribbed. **FLOWER** Sepals oval, lateral sepals asymmetrical with outer surfaces dark reddish-purple (thus unopened buds very dark) and inner surfaces greenish; petals paler and more strap-shaped; sepals and petals form a compact hood that embraces the column, reddish-purple when the flower is fresh although it quickly fades (and see var. *aestivalis* below). **LIP** White with a few scattered reddish-purple spots (formed by minute papillae), deeply lobed to form two broad chubby arms and two short stumpy legs, sometimes with a tiny tooth between them. A groove in the centre of lip towards the base leads into a short, conical, down-curved spur.
COLUMN Short and whitish; pollinia yellow.
SCENT The early-flowering variety has a strong, sweet, honey-like scent said to be similar to Heliotrope. Var. *aestivalis* has a weak, citron-like scent that is vaguely unpleasant. **SUBSPECIES** None.
VARIATION Very unusually there are two varieties differing in flowering time. The early-flowering **var. *ustulata*** is much the commoner. The late-flowering **var. *aestivalis*** is only found in S England. These two varieties are rarely if ever found together, but at a site in East Sussex a colony of 200 late-flowering plants merges with a much smaller group of early-flowering plants on a N-facing chalk slope.

The two varieties differ slightly. In var. *aestivalis* the hood averages darker reddish-purple and the colour does not fade once the flower has opened. Also a distinct rose-purple wash to the edges of the lip (may spread across the lip). Var. *ustulata* never shows the persistent dark hood or the coloured fringe to the lip. Other differences are rather subtler. On average, var. *aestivalis* is taller, with a more open spike of slightly smaller flowers, and the lip is also shorter, with a slightly narrower 'waist' and shorter legs which have a deeper cleft between them. Also, the spots on the lip are larger. Genetic analysis shows that there is far too small a difference between the early- and late-flowering varieties to justify treatment as separate species. **Var. *albiflora*** has unmarked white flowers. It is very rare but has been recorded in Derbyshire, Kent, Berks, Hants and Wilts. **HYBRIDS** None.

DENSE-FLOWERED ORCHID

Neotinea maculata

IDENTIFICATION

Confined to W Ireland. Height 6–15cm (4–30cm, very rarely to 40cm). Small and rather inconspicuous, and a hard orchid to spot, especially from walking height. The tiny off-white flowers are distinctive. **Similar species** Superficially similar to Small White Orchid, but the lip of the flower is a very different shape. **Flowering period** Late April–early June, usually peaking in mid May. The flowers go over very quickly.

HABITAT

Typically found in short turf on pastures, road verges, limestone pavements and around loughs and turloughs. Occasionally recorded on dunes or from ash and hazel woods in the hills. Although mostly confined to calcareous soils on limestone, has been found growing on gravels and also on light, peaty soils overlying more acidic rocks. Occurs up to 300m above sea level but most are found below 100m.

POLLINATION & REPRODUCTION

The spur contains traces of nectar, suggesting that the flowers are attractive to small insects. Although cross-pollination is possible, however, most or all plants are self-pollinated, sometimes even before the buds have opened. Whether self-pollinated or cross-pollinated, seed is produced in good quantities (*c.* 1350 seeds per capsule).

DEVELOPMENT & GROWTH

No specific information.

STATUS & CONSERVATION

Most populations are in The Burren region of Co. Clare and adjacent Co. Galway, including the Arran Islands. Otherwise a few very scattered sites in eastern Co. Cork, Co. Limerick and Co. Donegal and in Northern Ireland in Co. Fermanagh.

▲ Var. *alba*

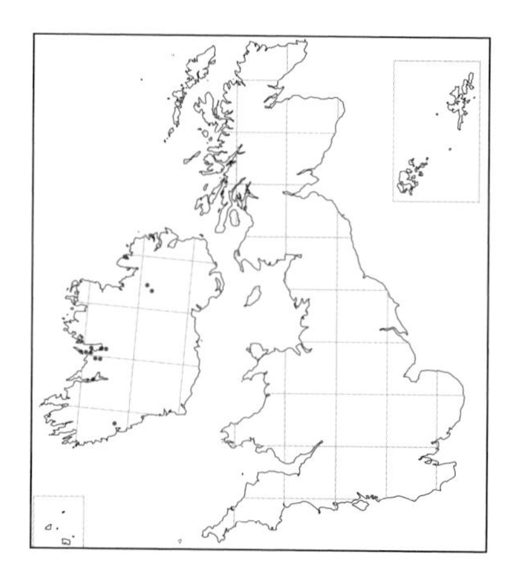

Scarce and local (although locally frequent in The Burren), usually found in small scattered colonies, less commonly as single isolated plants. The species has been lost from 46% of the historic range. Many of the losses are comparatively recent, although some of the decline has been offset by the discovery of new sites. The 'improvement' of pastures and overgrazing are likely causes for the losses. With the current rapid pace of development in Ireland there is cause for concern about the future of such a localised orchid.

Has a very peculiar world range, being essentially a Mediterranean species adapted to the mild wet winters and hot dry summers of that region and seemingly ill-fitted to the damp, windy, oceanic climate of W Ireland. It is not unique, however, as several other Mediterranean species, such as Strawberry-tree, occur in Ireland. Various theories have been advanced to explain this distribution, including land-bridges between Ireland and the Continent or glacial refuges off W Ireland, and the subject has generated a good deal of controversy. Colonisation of Ireland via wind-blown seed seems the simplest and most likely explanation.

Discovered on the Isle of Man in 1966 in an area of dunes on the N coast at The Ayres, Ballaghennie. This colony persisted until 1986 but Dense-flowered Orchid is now extinct there.

DESCRIPTION

VAR. *ALBA* STEM Green, with 1–2 brown membranous sheaths at extreme base. **LEAVES** 3–6 rather dark green leaves, of which 2–3 (–4) are broadly oblong-elliptical in shape and held rather spreading at base of stem; the remainder are narrower, more pointed and loosely sheathe the stem with the uppermost even smaller and more bract-like. Leaves appear in October and are wintergreen. **SPIKE** Cylindrical and dense, with 15–20 (–35) tiny flowers, all facing more-or-less the same direction. **BRACT** Green, becoming whitish towards tip, lanceolate, pointed and sometimes with a tooth at the side; a little shorter than ovary, which they clasp. **OVARY** Pale green, fat, cigar-shaped and twisted, with three diffuse ribs; rather larger than flower and held upright, it narrows and bends at the tip to hold the flower facing outwards. **FLOWER** Small and whitish, resembling a tiny man with a rather oversized helmet. Sepals white, washed greenish, with green veins; oval-lanceolate to lanceolate, pointed,

▼ Var. *alba*

◀ Var. *maculata*

fused at base to form a tight elongated hood. Petals greenish-white with green veins, very narrow, strap-shaped and pointed, and enclosed within hood. **Lip** Small in relation to hood, held at an angle forwards and downwards. Three-lobed, central lobe strap-shaped and notched or shallowly forked at tip (sometimes with a small tooth in notch). Side-lobes positioned near base of lip, rather shorter and narrower. Spur very short and conical. **Scent** Reported to smell faintly of vanilla. **Subspecies** None. **Variation** **Var. *alba*** has unspotted leaves and whitish or creamy flowers (rarely more greenish). This is the common variety in Ireland. **Var. *maculata*** has spotted leaves with reddish or purple spots arranged in parallel longitudinal lines. Lip pale pink or marked with a longitudinal pink stripe and sepals and petals with pink or brown veins. Occasionally the stem is spotted and sometimes also the bracts, ovaries and even the sepals and petals. This is the 'typical' variety (in that it takes its name from the species) but is relatively scarce in Ireland. **Var. *luteola*** has primrose-yellow flowers. It is very rare and perhaps extinct in Ireland. **Hybrids** None.

▼ Var. *alba*

PYRAMIDAL ORCHID *Anacamptis pyramidalis*

IDENTIFICATION

Widespread and locally common. Height 20–60cm (10–75cm). The dense spikes of unmarked bright cerise-pink flowers are distinctive. When the first flowers open, the spike forms a pyramid or cone, hence the name, but later it becomes more of a globe or cylinder. The flowers have a deeply 3-lobed lip and a very long, thin, down-curved spur. **SIMILAR SPECIES** Fragrant orchids have a looser, taller, thinner and more tapering flower spike, with slightly more purplish-pink flowers with the lip not so deeply cut. If there is any doubt, Pyramidal Orchid can always be identified by the *two prominent raised ridges* or 'guide-plates' at the base of the lip, unique to the species. **FLOWERING PERIOD** Early June–mid August, very exceptionally from May, but mostly mid June–mid July.

HABITAT

Dry, well-drained grassland on chalk, limestone or other calcium-rich soils, such as boulder clay; indeed, almost anywhere with a hint of lime in the soil. Found in both close-cropped turf and in taller, ranker swards. Suitable habitats also include the grykes of limestone pavements, cliff tops, amongst Marram on sand dunes and, very locally, the coastal machair on the Hebrides. Sometimes still found in the few surviving old meadows. Also grows among scrub and rarely in open woodland with a broken canopy. Takes readily to man-made habitats, such as road verges and roundabouts, churchyards, old quarries, disused railway lines and old industrial sites.

POLLINATION & REPRODUCTION

Pollinated by moths, both day- and night-flying, and by butterflies. Day-flying moths and butterflies are attracted by the flower's vivid coloration, and the various burnet moths may additionally be attracted by its similarity to the bright red spots on their wings. At night the flower's scent would assist its detection.

The pollination mechanism is highly evolved. The walls of the spur contain a sugary sap but only an insect with a suitably long proboscis can reach this. The ridges on either side of the mouth of the spur act as a guide to correctly to position the insect's proboscis, not only to access the spur and the sap within but also to trigger the mechanism. The rostellum hangs over the entrance to the spur and when touched by an insect the protective, flap-like bursicle is pushed aside and the strap-shaped viscidium, complete with the two pollinia, sticks to the insect's proboscis. As soon as the viscidium is exposed to the air it contracts like a watch-spring and coils itself around the proboscis. This not only helps to attach it more firmly but also moves the pollinia apart so that they become separated by 90°. Then, a few seconds later in a second contraction, the pollinia swing forward so as to be in a perfect position to strike the sticky surfaces of the two stigmas on the next flower visited and deposit packets of pollen. The mechanism is efficient, with 65–95% of flowers setting seed.

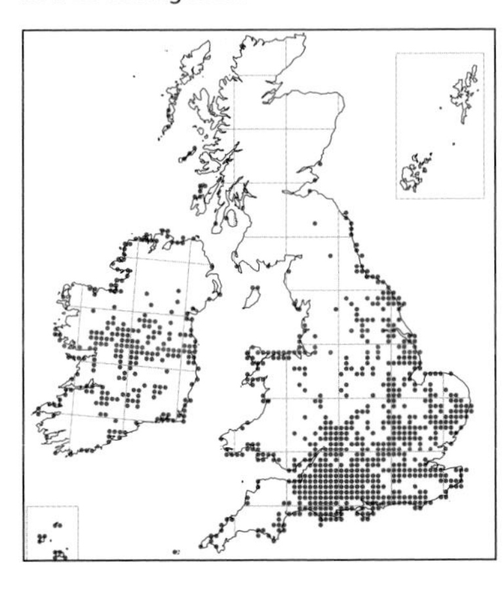

Vegetative propagation is also possible via the production of additional tubers, which often develop at the end of short rhizomes.

DEVELOPMENT & GROWTH

The tuber's roots are infected with fungi. The basal leaves appear in autumn and die down in summer and after flowering the orchid spends the late summer and autumn 'resting' as a tuber.

It is not known when germination occurs but initially a protocorm is developed. It may be just a few months before the first tuber develops. Following this the protocorm dies off in the late summer to leave just the small tuber in a 'resting' state. In the autumn a bud on the tuber produces a short rhizome and from this one or two roots develop and sometimes also the first leafy shoot. The new rhizome and roots have to be reinfected with fungi from the soil as the tuber itself did not carry any infection. This plant, including the leafy shoot, overwinters, and in the following spring a new tuber starts to develop from a bud on the rhizome and the old one starts to shrivel away. Eventually the shoot and rhizome also wither, and the plant will again spend the late summer 'resting' as a tuber. In this way the annual cycle of growth is established. The period between germination and flowering could be as little as three years.

STATUS & CONSERVATION

The generally southerly distribution, with colonies in the N concentrated near the coast, emphasises its preference for a mild winter climate. Rare in Scotland.

The boundaries of the range are stable but inevitably there have been declines; lost from 20% of the historical range in Britain and 31% in Ireland. Some of the decline is due to the loss of old meadows and pastures. Although tolerant of grazing, overgrazing may be a threat, perhaps especially at the few sites on the machair of the Hebrides and W Scotland. Another

issue is the development of old industrial sites and old pits and quarries.

Until recently the Pyramidal Orchid was the only member of the genus *Anacamptis*, but it has now been joined by Green-winged and Loose-flowered Orchids.

DESCRIPTION

Underground The aerial stem grows from a pair of rounded or roughly elongated tubers with a few slender fleshy roots growing almost horizontally near the soils's surface. **Stem** Green, slightly angled towards tip and often rather slender and flexuous: stems frequently have a distinct kink; 2–3 brown sheaths at the extreme base. **Leaves** Green, tinged grey, keeled and strap-shaped with a pointed tip; 3-4 sheathing leaves grade into 5-6 non-sheathing leaves that decrease in size up the stem and become bract-like towards spike. Sheathing leaves emerge in autumn and are wintergreen, but have often withered by flowering time. The spike and non-sheathing leaves do not appear until spring. **Spike** Very dense, with 30–100 flowers packed so closely that they often conceal the stem. **BRACT** Green, sometimes flushed purple, narrow, a little longer than ovary and tapering to a fine point. **OVARY** Cylindrical, twisted and green, often washed reddish-purple. **FLOWER** Usually vivid pink, sometimes paler pink or reddish-pink, fading a little as the flower ages. Sepals oval-lanceolate with a variably pointed tip and dished sides, petals a little shorter and blunter. Lateral sepals held horizontally or slightly drooping, to about 15° below the horizontal; upper sepal and petals form a hood over the column. **LIP** White towards base, wedge-shaped, held projecting forwards and downwards. Deeply divided into three virtually equal strap-shaped lobes; outer lobes broaden slightly towards their blunt, shovel-like tips, central lobe more parallel-sided with extreme tip pinched in and upwards. On either side of base of lip two prominent narrow raised ridges, extensions of the column; although roughly parallel, they converge towards mouth of spur. Spur slender, 12–14mm long (often longer than ovary) and down-curved. **COLUMN** White, variably washed pink. The rostellum partially blocks entrance to spur; two stigmas lie low down on either side of column; both pollinia are joined to the same strap-shaped viscidium. **SCENT** Most obvious in the evening, variously described as sweet or as slightly unpleasant and 'foxy'. **SUBSPECIES** None. **VARIATION Var.** ***albiflora*** has pure white flowers but is rare (very pale pink flowers are rather commoner). **Var.** ***sanguinea*** has blood-red flowers. Scarce, known only from the Hebrides and NW Ireland. **Var.** ***emarginata*** has an unlobed lip resembling a scallop shell. Very rare. **Var.** ***angustiloba*** has a very deeply lobed lip. **Var.** ***fundayensis*** has a taller and more cylindrical flower spike. Described from Funday in the Hebrides but probably extinct, if it ever existed. **INTER-GENERIC HYBRIDS X** ***Gymnanacamptis anacamptis***, the hybrid with Chalk Fragrant Orchid, has been reported once each in Hants and Co. Durham.

GREEN-WINGED ORCHID *Anacamptis morio*

IDENTIFICATION

Scarce and very local, but can occur in large numbers where found. Height 7.5–15cm (5–30cm, rarely to 50cm). A dainty, usually petite orchid, with unspotted leaves and a few, relatively large flowers extremely variable in colour, from deep violet-purple to rose-pink or whitish, although most are a shade of purple. All 3 sepals and the 2 petals form a hood marked with *fine green or bronze veins* that give the species its name, and the long straight spur projects conspicuously back from the flower. **Similar species** Early Purple Orchid is superficially similar but usually has spotted leaves and its flowers always lack obvious dark veins on the sepals, while only the upper sepal and petals form the 'hood'; the two lateral sepals are held upright as 'wings'. **Flowering period** Mid April–mid June but mostly in May. Earliest in the W and exceptionally flowers in mid March.

HABITAT

Unimproved grassland where grazing, mowing or other factors keep the grass relatively short and the sward open. Strongly associated with old, species-rich grassland and slow to colonise new sites. The optimum habitats are damp pastures on clay soils, but also found in dry chalk or limestone grassland and sometimes on sands and gravels, including pockets of neutral grassland on otherwise acidic heathland. Suitable meadows and pastures are now rare. Also found on more marginal sites such as stabilised dunes, lime-rich eskers, old railway cuttings and banks, village greens (especially 'commons'), churchyards and golf courses. Occasionally found on lawns or old industrial sites. Intolerant of shade and, although sometimes found amongst scrub, seldom if ever found in woodland. Recorded up to 305m above sea level (Co. Roscommon).

POLLINATION & REPRODUCTION

Pollinated by bees, especially bumblebees, but produces no nectar and relies on deceit. When queen bees first emerge from hibernation in the spring they are 'naive' and have yet to learn which flowers offer a genuine reward of nectar; they are easily attracted by the brightly coloured orchids. Bees usually visit a single flower on each of several plants before they realise there is no reward and move on to more profitable

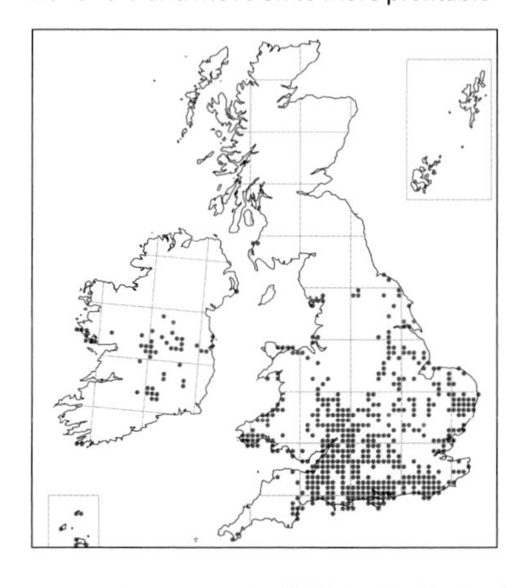

species. On each orchid visited, they start with the lower flowers and therefore these are the ones that tend to set seed. A cold spell in May, which disrupts the bees' routine, benefits the orchid because the bees have to relearn their foraging routes once the weather warms up again. Pollination is variably efficient; in a Swedish study only 5–30% of flowers set seed but in Britain almost all flowers may be pollinated. Each capsule produces 4,000 seeds. May also reproduce vegetatively via the formation of additional tubers.

DEVELOPMENT & GROWTH

Probably long-lived: in a study in Cambs some plants flowered for 17 out of 18 years and almost all flowered several times, interspersed with seasons in which they just produced a rosette of leaves or more rarely were dormant underground. At Iron Latch Meadow in Essex, Green-winged Orchids reappeared and flowered after an absence of *c.* 30 years when closed-canopy hawthorn scrub was removed.

Seeds are thought to germinate in the late summer or autumn, and the protocorm probably produces the first roots and even the first small leafy shoot during its first winter or spring. The plant goes on to develop its first tuber at the base of the leafy shoot, and it is this tuber alone that will persist through the late summer 'resting period', establishing the annual cycle. It is thought that plants can flower within three years of germination; in cultivation they have flowered after two.

STATUS & CONSERVATION

Listed as Near Threatened. Once common and widespread, it has been lost from many sites and, although thousands still grow in favoured locations (mostly reserves and SSSIs), has overall declined dramatically in the last 75 years. Now gone from at least 49% of the historical range in Britain and 60% in Ireland. The decline is almost entirely due to agricultural changes. In the 19th century its favoured damp meadows and pastures were ploughed and converted to arable. In the 20th century, especially since 1945, permanent pastures have disappeared from many farms and those that survive have been drained, ploughed and reseeded. Research has shown that Green-winged Orchid will decline significantly if fertilisers are applied to grassland, probably due to competition with the increasingly vigorous grass and herbs, and the phosphorus in fertilisers may actually be toxic to the orchids. If anything, habitat losses have accelerated in the last 30 years as the last few pristine pastures have been 'improved'; the decline in the amount of suitable habitat must approach 99%.

DESCRIPTION

STEM Yellowish-green, washed purple towards tip and slightly angled, with 2–3 thin whitish sheaths at the extreme base. **LEAVES** Up to seven bluish-green basal leaves, varying from roughly elliptical to lanceolate, keeled, held semi-erect; 2–3 more pointed leaves higher on stem, becoming bract-like below spike. Leaves wintergreen, emerging in September-October and dying down by mid June. **SPIKE** Rather open and loose with 4–14 well-spaced flowers. **BRACT** Green, typically washed purple, lanceolate, 2/3–2x length of ovary, which they curl around and sheathe. **OVARY** Green, variably tinged purple, cylindrical, strongly curved (through *c.* 90°), ribbed and twisted. **FLOWER** Various shades of purple, although a few are rose-pink or whitish. Sepals oblong-oval, upper sepal a little narrower than lateral sepals, generally whitish with a variable violet-purple wash, deepest on upper sepal and upper margins of lateral sepals, which have 3–7 bold, green or bronze parallel veins visible on both surfaces. Petals coloured as upper sepal but shorter, narrower and more strap-shaped. Petals and sepals form a 'hood' enclosing the column. **LIP** Rather broader than long with three lobes; side-lobes rather larger than central lobe, rounded, often with crinkled (crenate) edges and folded downwards; central lobe roughly equal in length to side-lobes. Typically violet-purple, whiter in centre and at mouth of spur, usually with some violet-purple blotches or spots in paler central area. Spur coloured as lip, *c.* 1/2 length of ovary, narrow but flattened towards tip, where often notched and slightly curved upwards. **COLUMN** Whitish, washed purple around pollinia. **SCENT** At least some have a vanilla-like scent; white-flowered forms may, on average, have a stronger perfume; some people cannot smell it at all. **SUBSPECIES** None. **VARIATION Var. *alba*** has white flowers, sometimes very faintly tinged pink, but still with green veins. Usually scarce (0.1–1% of plants) but sometimes as much as 15% of a population. **Var. *bartlettii*** is small with very small flowers, **var. *churchillii*** is tall; neither seems well defined. **INTER-GENERIC HYBRIDS X *Anacamptorchis morioides*,** the hybrid with Early Purple Orchid, has been recorded rarely and sporadically in England and Wales.

LIZARD ORCHID *Himantoglossum hircinum*

IDENTIFICATION

Rare and very local, but prone to turn up unexpectedly in new places. Height 25–60cm (–100cm), commonly 30–45cm. The tall spikes of greyish-green flowers have an untidy, ragged appearance, and the flowers are unique. The long central lobe of the lip resembles a lizard's tail, the shorter side-lobes forming the back legs. The hood of the flower is said to recall the head and body of the lizard, but it appears to me that the fore-quarters of the lizard have been swallowed and have vanished into the throat of the flower. **Similar species** None. **Flowering period** Early June–late July, sometimes from late May. Often at its best mid–late June.

HABITAT

Open sunny places on shallow, well-drained soils, usually on chalk, limestone or coastal dunes rich in shell-sand, occasionally on boulder clay, more acid sands or gravels and even recorded growing through broken tarmac. Often in fairly rank grass, although sometimes also in shorter swards, and can appear amongst scrub or on woodland edges. Suitable sites include road verges, railway embankments, ancient earthworks, field margins, old chalk pits and even lawns. At least six of the 19 recent populations are on golf courses.

POLLINATION & REPRODUCTION

The flowers are visited by a variety of insects, but solitary bees of the genus *Andrena* are probably the main pollinators. There is no nectar and the flower offers no reward; its scent may serve to attract flies and night-flying moths. Self-compatible – solitary isolated plants can set seed, most probably by cross-pollination from other flowers on the same spike (geitonogamy, which may account for *c.* 30–35% of seed produced in all populations). Whatever the mechanism, it does not seem to be very efficient and only *c.* 30% of flowers are pollinated. Each capsule contains up to 1,200 seeds, perhaps more. May reproduce vegetatively by forming extra tubers, but this seems to be a rare event.

DEVELOPMENT & GROWTH

Spends the summer 'resting period' underground as a tuber. The aerial shoot appears from late August, with the lower leaves unfolding and some short, thick roots growing from the base of the stem. The leaves overwinter and by early spring may start to blacken at the tips. By then a new tuber has started to form from a bud at the base of the stem; by May the roots and older tuber, as well as the leaves, are often withering. At flowering time, therefore, has two large, egg-shaped tubers (one old and withering, one new). After flowering and setting seed the plant will disappear, leaving only the new underground tuber to survive the summer.

Individual plants can be long-lived, surviving for up to at least 19 years after first emergence, but may not flower every year and can remain dormant underground for a year or appear merely as a rosette of leaves, sometimes for many years. An individual plant in Sussex

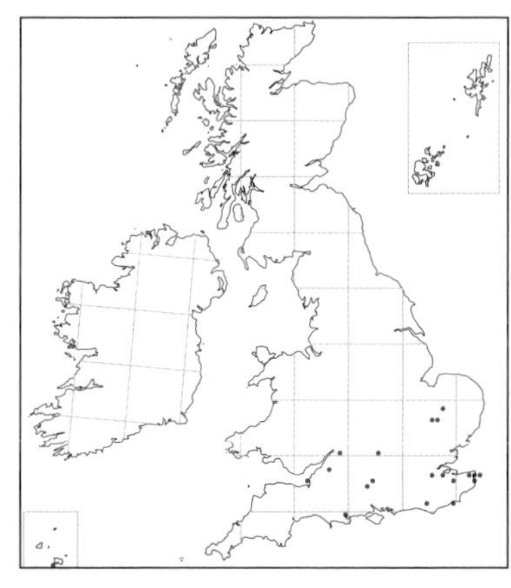

flowered in 1984 but not again until 1995, and a Suffolk colony of 16–40 plants did not produce any flowers over ten years. Wet autumns and warm, wet winters free from sharp frost encourage flowering, while hard winters and spring drought may lead to flowering being aborted.

Seed ripens 6–8 weeks after pollination and often germinates immediately (although seed may be viable for three years). With the aid of fungi, the seedling develops into a protocorm the size of a small pea. The following spring this protocorm produces the first tuber and then withers away, leaving the tuber to oversummer. In the autumn, a short rhizome grows from this tuber and develops a root and a second tuber that is fully developed by the following spring. This establishes the annual cycle of replacement tubers. When the seedling is 2–3 years old the first leafy shoot appears above ground. The seedling usually remains in the one-leaf stage for at least two years and then moves on to two leaves, again for several years, and then three leaves. Although in very favourable conditions a plant with 2–3 leaves can flower, it is only when it has grown four leaves that it can be considered mature. Even then it may not flower every year. The minimum period between germination and first flowering may thus be six years.

STATUS & CONSERVATION

Nationally Scarce and listed as Near Threatened: WCA Schedule 8. On the edge of its range in England, subtle shifts in climate may have a big impact on its distribution, which has ebbed and flowed. From its discovery until *c.* 1850 Lizard Orchid was restricted to the Dartford area of Kent. It then declined with just isolated records in SE England, mostly in Kent; by 1900 there were only four sites and it was thought to be near extinction. Then, from *c.* 1915, there was a marked expansion northwards and it turned up unpredictably, often just a single plant but sometimes in larger numbers, N to Yorkshire and W to Devon, peaking *c.* 1927 with 36 populations. There was then a sudden drop in the number of colonies after 1934, although the geographical spread was maintained. From the mid 1940s–1993 the number of populations stabilised at 9–11, although some sites were lost and others gained. During this period all colonies were small and even the most thriving lasted only *c.* 20 years.

A further increase then started *c.* 1994 with plants appearing W to Somerset and Glos. There were 16 sites in 1996 and 19 by 2000, including a very large colony at Sandwich Bay in Kent, with up to 3,000 flower spikes (5,000 in 2000, when the total population was estimated at 27,500 plants in 0.5km^2) and a smaller permanent colony in Suffolk on the Devil's Dyke on Newmarket's July racecourse, with 200–250 flowering plants.

Favours areas of sparse long grass cut once or twice a year. No sites are grazed but the species is tolerant of grazing, at least by cattle; rabbits have caused severe damage at times, especially during hard winters, as have slugs. Some sites are managed to open up the vegetation and prevent scrub invasion (with the Newmarket site, for example, being burnt every fifth year during the winter).

DESCRIPTION

Underground The aerial stem grows from two oval tubers. **Stem** Pale green with faint purple blotches, ridged towards tip; 2–3 scale leaves at extreme base. **Leaves** Greyish-green. 4–10 large, oval-oblong, keeled basal leaves form an untidy rosette; 3–5 smaller, narrower and more pointed leaves loosely clasp the stem up to the flower spike. **Spike** Occupies *c.* 30–50% of stem, cylindrical, rather dense, with 15–80 (–200) flowers. **Bract** Pale green to off-white, variably but often strongly washed reddish, narrow and pointed, up to 2x length of ovary. **Ovary** Pale green, cylindrical, twisted and 3-ridged, tapering at the base into a short stalk. **Flower** Sepals oval, fused at base to form a loose hood; pale greyish-green, variably flushed or rimmed purple with paler interior with lines of reddish-purple spots and dashes along veins (may show through to exterior). Petals very narrow, strap-shaped, spotted reddish-purple. **Lip** 3-lobed, central lobe 2.5–6cm long, narrow, parallel-sided and ribbon-like, usually with deeply notched tip, twisted through 2–3 turns, whitish at base, becoming pale lilac-brown for outer 80%; side-lobes very much shorter and a little narrower, pointed, pale lilac-brown. Sides of lip at base boldly folded or corrugated, this corrugation continues on to outer edges of side-lobes. Centre of base of lip trough-like with a dense 'fur' of tiny white papillae, with variable bright purplish-red tufted spots and blotches. Lip coiled like a spring in bud and slowly unrolls to project outwards and downwards at *c.* 45°. Spur short, curved down and bluntly conical. **Column** Anther greenish-white; two greenish pollinia fixed to a single viscidium concealed in purple bursicle. Stigmatic zone purple. **Scent** Musty, usually said to recall a billy goat and most pungent in the evening (but not always evident). **Subspecies** None. **Variation** None. **Hybrids** None.

FLY ORCHID *Ophrys insectifera*

IDENTIFICATION

Rather local but inconspicuous and easy to overlook. Height 15–60cm. The tall slender spikes with small, well-spaced flowers can be very hard to see among other vegetation and even on a bare woodland floor can vanish with ease. Very distinctive, the individual flowers are indeed like little flies and are the most wonderful example of insect-mimicry amongst British orchids. The purplish-brown lip forms the 'body' and the lustrous slate-blue speculum shines like folded 'wings'. The two glistening depressions at the base of the lip are the 'eyes' and above these the dark, wire-like petals look just like little antennae. **SIMILAR SPECIES** None. **FLOWERING PERIOD** Late April–early July but mostly late May–early June.

HABITAT

Very varied, although usually grows on calcareous soils over chalk and limestone. Found in open deciduous woodland and in S England particularly associated with beechwoods. Favours better-lit areas in glades, rides and along the edge of woodland, as well as shaded road banks and open scrub, but sometimes grows in deeper shade such as overgrown hazel coppice. In S England occasionally found on open grassland in old pits, quarries and on spoil heaps, but when it does grow in such habitats it may occur in large numbers; has been recorded rarely from slumped coastal cliffs. In N England and Ireland probably more frequent in open areas and in addition to wooded sites found on limestone pavements and rocky hillsides. Grows in alkaline fens and on the margins of turloughs in W Ireland and also in fens on Anglesey, often among tussocks of Black Bog-rush. Recorded up to 390m above sea level (Cumbria).

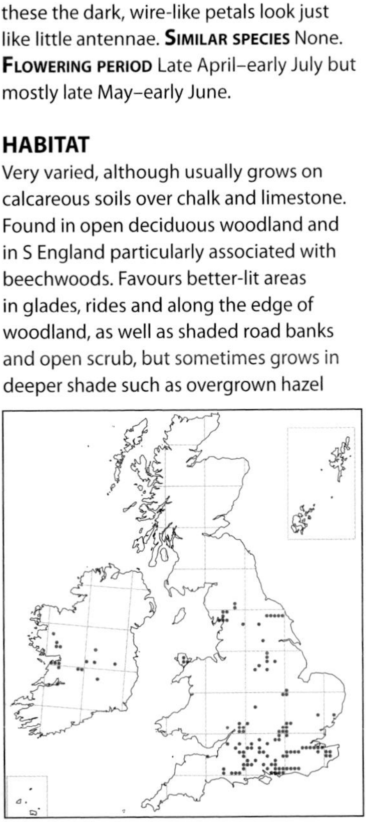

POLLINATION & REPRODUCTION

Pollinated by male digger wasps *Argogorytes mystaceus* and also *A. fargeii*, a species that is rarer and emerges a little later in the season. The wasps are attracted by pheromones emitted by the orchid and by the shape and texture of the flower. They attempt to copulate with it and during this 'pseudocopulation' the pollinia are attached to their heads. Pollination rates are variable but often low: rates of 2.1% and 7.5% have been quoted for two samples of *c.* 1,000 plants, while a much smaller sample from Somerset achieved a rate of 35%.

DEVELOPMENT & GROWTH

The first leaf appears in the winter after germination and the first tuber is formed in the second year. In mature plants leaves emerge in autumn and are wintergreen.

STATUS & CONSERVATION

Listed as Vulnerable. Widespread but very local. There have been considerable losses and now occupies only 42% of its total historical range in Britain. Much of the decline took place long ago, especially in East Anglia. However, in Ireland, where now found in just over 50% of its historical range, rather more of the losses have been recent. Causes include woodland clearances and 'coniferisation' but perhaps equally important has been the maturation of woodland and scrub due to changes in forest management or its abandonment altogether. In such cases increasing shade probably means that the flowers are seldom pollinated (the woods are too shaded for the pollinating wasps to thrive) and the population declines and eventually disappears.

DESCRIPTION

UNDERGROUND The aerial stem grows from a pair of tubers, the younger of which is often stalked. **STEM** Pale green, slender, with 1–2 basal sheaths. Groups of up to ten plants may grow together. **LEAVES** 2–5, shiny, dark green or bluish-green, the lower narrow and strap-shaped, flaccid but keeled and usually pointed at tip, upper 1–2 narrower, more pointed and loosely sheathing stem. **SPIKE** Although initially bunched, the 1–10 flowers (exceptionally –20) are well-spaced along the stem by the time the uppermost has opened. **BRACT** Dark green or bluish-green, lanceolate, often with edges rolled inwards; lower bracts rather longer than ovary but towards tip of spike they are a little shorter. **OVARY** Pale green, slender, cylindrical, 6-ribbed; upright but curving at the tip to hold the flower facing outwards. **FLOWER** Sepals yellowish-green, oval-oblong with a blunt tip but appear narrower because their edges are rolled back and the inner face is concave. Lateral sepals held horizontally, upper sepal vertical but arching forward over column to a variable extent. Petals dark purplish-brown with short, fine, downy hairs; much smaller than sepals (less than half as long); their edges are rolled back to give them a fine, filiform appearance and they point forward. **LIP** Longer than wide, hanging down almost vertically, divided into three lobes: two relatively short, narrow side-lobes spreading outwards at base and a broad terminal lobe notched or forked at tip. Lip velvety in texture (the side-lobes are hairier), rich dark reddish-brown or

purplish-brown, becoming a little paler towards tip and duller with age. A more-or-less square, shining, pale slate-blue band across the centre forms the speculum; at base of lip are two shining 'pseudoeyes'. **COLUMN** Short, reddish-brown with small, circular stigmatic cavity at base and short, blunt 'beak'. Pollinia two, yellow; viscidia enclosed in two off-white bursicles. **SUBSPECIES** None. **VARIATION** Considerable variation in the colour and markings of the lip. In **var. *ochroleuca*** the lip is pale yellowish-green with a white speculum; recorded from Kent, Hants, Wilts and Herts. **Var. *flavescens*** is a little darker, with a yellowish-brown lip, pale blue or whitish speculum and greenish-brown petals; recorded from Glos. **Var. *subbombifera*** has a very broad, rounded central lobe, notched as usual at tip. Very rare, recorded from Surrey and Hants. **Var. *parviflora*** has flowers *around half* normal size. Very rare. **Var. *luteomarginata*** has a broad yellow border to the central lobe, often also a yellow stain or tip to side-lobes and yellow or green tips to petals; recorded rarely in Surrey, Hants, Glos and Anglesey. **HYBRIDS *O. x pietzschii***, the hybrid with Bee Orchid is very rare; recorded in West Sussex and at two sites N Somerset. ***O. x hybrida***, the hybrid with Early Spider Orchid, has occurred occasionally in E Kent.

EARLY SPIDER ORCHID *Ophrys sphegodes*

IDENTIFICATION
Almost entirely restricted to the S coast between Dorset and Kent, where very locally abundant. Height 5–15cm (–20cm, rarely 35cm, but at Samphire Hoe plants may be as tall as 45cm). Unmistakably a 'bee' orchid, the circular deep brown lip resembles the body of a garden spider and is marked with a lustrous bluish speculum. Sepals green, petals yellowish-green, narrow and strap-shaped with wavy edges. The flowers fade rapidly, the lip becoming a dull pale yellowish-brown or grey-brown. **Similar species** Bee and Late Spider Orchids have similar large brownish lips but their specula are bordered by a narrow creamy line and their sepals and petals are pink. **Flowering period** Late March–early June. Tends to flower earliest in Dorset (early–mid April to early–mid May) and may average a few days later in Sussex and Kent. Flowering times are very variable, however, both from year to year and between colonies, even in the same area.

HABITAT
Most sites are on old, species-rich grassland on chalk or limestone, both in short closely-cropped turf (even lawns) and in slightly ranker swards. Has some preference for previously disturbed areas, such as old quarries, spoil heaps and tracks; as with Late Spider Orchid heavy ground disturbance may aid the successful establishment of new colonies. It has been recorded rarely from shingle. Almost all the current sites are near the sea.

POLLINATION & REPRODUCTION
Pollinated by male solitary bees of the species *Andrena nigroaenea*, initially attracted by a scent that mimics the bee's pheremones, then by sight and finally by touch: the pollinia are attached to the front of the bee's head as it attempts 'pseudo copulation'. The mechanism can be successful with *c.* 25% of flowers recorded as setting seed at Samphire Hoe, but pollination rates can be very much lower, with a lack of suitable pollinators being the most likely explanation for the low seed-set. It has also been suggested that British populations are probably mostly self-pollinated. Indeed, the pollinia are occasionally released to dangle in front of the stigma as in Bee Orchid, making self-pollination possible. The overall low rates of pollination suggest, however, that self-pollination is not routine. Vegetative reproduction is thought to be uncommon or rare and as Early Spider Orchid is a short-lived orchid, dependent on seed to maintain its numbers, sufficient seed is presumably being produced to sustain the current colonies despite the apparently low rates of pollination.

DEVELOPMENT & GROWTH
No information on the period between germination and first appearance above ground, but probably *c.* 1–3 years. First appeared at Samphire Hoe four years after spoil from the Channel Tunnel was spread.

Short-lived. Few plants live for more than three years after their first emergence above ground and the majority appear

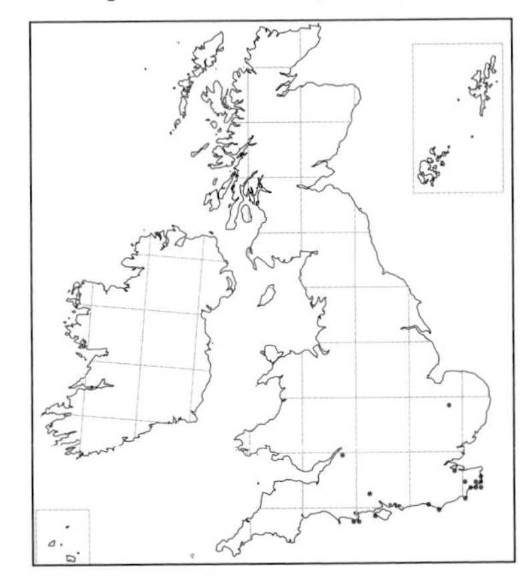

just once, flower and then die (i.e. they are monocarpic); a tiny minority can live for at least ten years. The number of plants above ground varies widely from year to year, largely correlated with the amount of rainfall the previous winter; plants may spend 1–2 years dormant underground.

In a study at Castle Hill, Sussex, over a winter season, plants began to appear above ground in early September and continued to emerge through the winter, with peaks in November–December and more especially March–May. By flowering time therefore some plants had been above ground for two months and some for six. Those that appeared early in the winter included most of the older plants that had flowered the year before and all these early plants suffered grazing damage. Previously unrecorded plants, presumably seedlings emerging above ground for the first time, appeared throughout the season but especially from March onwards. However, *c.* 75% of all the plants recorded up to March did not survive above ground to be counted in the annual census in May, with grazing being the likely cause of their disappearance.

STATUS & CONSERVATION

Nationally Scarce: WCA Schedule 8. Largely confined to the coasts of Dorset, E Sussex and Kent; in Dorset there are large populations between St Aldhelm's Head and Durlston (estimates of *c.* 50,000 plants); in E Sussex now found mostly between Beachy Head and Seaford and at Castle Hill near Brighton; in Kent *c.* 30 colonies, most along the coast but including one or two inland sites. One of the best and most accessible sites is Samphire Hoe below Shakespeare Cliff at Dover, created using 5,000,000 m^3 of marl dug from beneath the seabed during the construction of the Channel Tunnel. In 1998 there were 61 plants but by 2004 this has increased to 9,000. Away from these strongholds there have been isolated recent records on the Isle of Wight (2013), in W Glos (from 1975), S Wilts (1988), Northants (2001, the first record for 230 years) and W Suffolk (where a single plant appeared from 1991 at Lakenheath, the first record since 1793). Introduced near Bishop's Stortford in Herts (from *c.* 1972).

Aside from its impressive populations on the S Coast, Early Spider Orchid has vanished from at least 73% of its historical range, but the majority of the losses occurred long ago, in the 19th century. This was probably largely due to the cultivation of grasslands following the Enclosures, although the retreat towards the coast of S England suggests that climatic factors may also have been involved. Many sites are now protected but its grassland habitat needs careful management to maintain populations. Grazing by sheep is beneficial, as long as they are removed during the period when the orchids flower and set seed.

DESCRIPTION

UNDERGROUND The aerial stem grows from a pair of tubers, the younger often stalked. **STEM** Yellowish-green, thick and fleshy. **LEAVES** Grey-green to green, prominently veined; 3–4 lower leaves short, blunt, broadly strap-shaped, the lowest nearly horizontal but the remainder more upright. Upper leaves narrower, more pointed and loosely sheathing stem. Some wintergreen but in most plants leaves appear in early spring; all have withered by the end of June. **SPIKE** Most have 2–7 (–10) flowers, but at Samphire Hoe some tall plants may have as many as 17 flowers widely spaced along the stem. **BRACT** Pale green, lanceolate, blunt-tipped, slightly longer than ovary, which they more-or-less sheathe at their base. **OVARY** Green, boldly 6-ribbed. **FLOWER** Sepals oval-oblong with blunt tip, margins rolled under; usually pale yellowish-green, rarely more whitish. Petals more strap-shaped, shorter and narrower, tip squared-off, edges frequently waved or undulate; often a slightly deeper yellowish-green, their margins may be more brownish with some fine hairs; petals very rarely pinkish. Lateral sepals held horizontally, upper sepal vertical, usually arching forward over column. **LIP** Almost circular, edges moulded downwards to give a convex profile. Deep purplish-brown, usually paler and more yellowish around lower edge, velvety in texture with a smooth, slightly lustrous, slate-grey, lead-coloured or bluish mark (the speculum) in centre or towards base, often in the form of an H, X or the Greek letter π. Two roughly conical swellings or haunches, one either side of base of lip, variable in size; lip particularly hairy around these and along the edges below them. **COLUMN** Greenish-white, held at *c.* 90° to the lip, the rostellum at its tip forming a short blunt beak. Stigmatic cavity at base of column circular, maroon-brown on lower half and pale green on upper with two 'pseudo-eyes', one on each side; it has been reported to contain 'sugar'. Pollinia yellow. **SUBSPECIES** None. **VARIATION** Shape and coloration of lip very variable; rarely the speculum may be red rather than blue. **Var. *flavescens*** lacks anthocyanin pigments: lip greenish or golden, sometimes browner towards edges, with a 'shadow' speculum. Rare, and easily confused with a normal flower that has faded with age. **HYBRIDS *O.* x *hybrida***, the hybrid with Fly Orchid, has occurred occasionally in E Kent. ***O.* x *obscura***, the hybrid with Late Spider Orchid, was recorded in Kent in 1828.

Var. *flavescens* ➤

BEE ORCHID *Ophrys apifera*

IDENTIFICATION

Widespread in England and parts of Wales and Ireland, in a wide variety of open grassland sites. Height 10–45cm (–65cm). Well named, the flower looks very much like a bumblebee and it has indeed evolved to attract male bees as pollinators (but see below). The lip of the flower forms the bee's body – velvety maroon-brown with a pattern of creamy markings and noticeably hairy, rounded side-lobes – and the slender, parallel-sided, greenish or pinkish-brown petals form two antennae. **SIMILAR SPECIES** In most regions the only 'bee' orchid and therefore distinctive, but in S England care should be taken to distinguish it from Early Spider Orchid (p.234) and, in E Kent, Late Spider Orchid (p.244). **FLOWERING PERIOD** Early June–late July, sometimes from late May, but at least in S England most plants will have finished flowering by the end of June.

HABITAT

Essentially grassland, from closely cropped swards to areas with much bare ground or ranker grassland among scrub. Most sites are on light, well-drained soils low in nutrients, but also grows on heavy clays, in areas that may have standing water in the winter, or on flushed, slumped clay cliffs. Usually stated to favour calcium-rich soils overlying chalk and limestone as well as chalky boulder clay, but actually much more tolerant and any poor, free-draining soil that is not too acid may be suitable. Many habitats are man-made, such as road verges, railway embankments and cuttings, old quarries, pits, spoil-heaps, gravel pits, brownfield industrial sites and garden lawns. Also found in more natural areas: sand dunes, dune slacks, limestone pavement and, especially in Ireland, fens. Tolerant of heavy grazing and trampling. Bee Orchids often behave like 'weeds', quickly colonising areas of bare or disturbed ground. They can increase rapidly in numbers until a dense, closed sward develops and then, being poor competitors, frequently disappear. On the other hand, they can sometimes thrive in permanent grassland and such colonies may last for many years.

POLLINATION & REPRODUCTION

Unlike all other members of the genus *Ophrys*, usually self-pollinated. Shortly after the flower opens the anther releases the pollinia, which have long, flexible, thread-like stalks and drop down to dangle in front of the stigma. In this position the slightest breeze will waft them on to its sticky surface and pollination will take place. This mechanism is efficient and a large proportion of flowers are pollinated. Each pod is estimated to hold 6,000-10,000 seeds. Much more rarely, the flowers may be cross-pollinated. 'Pseudo-copulation' involving bees has been recorded and Bee Orchid has hybridised with several other members of the genus *Ophrys* (only possible if it is cross-pollinated).

Vegetative propagation may occur; in cultivation several small 'daughter' tubers have been found at the end of long, slender rhizomes.

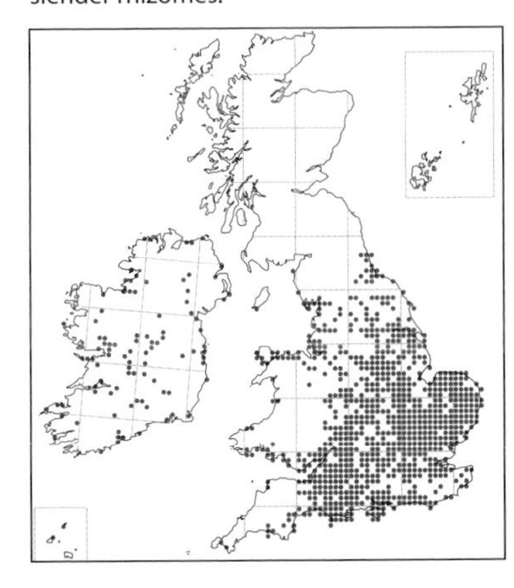

Seed probably germinates in the spring but there is no clear picture of the period between germination and the first appearance above ground. A 19th century author noted that germinating seeds, protocorms and plantlets with small leafy shoots were found around adult plants in March. Young Bee Orchids often appear above ground in the late winter as non-flowering plants with a single leaf but these frequently wither and die off rather quickly.

DEVELOPMENT & GROWTH

Relatively long-lived, some plants live for at least ten years after their first appearance and may flower for three years in a row. They can, however, spend 1–2 years dormant underground but such plants do not flower in the first season after they next emerge. They seem to need at least a year's growth as a non-flowering rosette to build-up enough reserves to bloom again. At any stage in their life they can only flower if they have reached a minimum leaf area, and if they do not have a big enough spread of leaves the rosette will wither early, in May.

STATUS & CONSERVATION

Widespread in England N to Cumbria and Co. Durham, although sparse or absent in N Devon and Cornwall, and the number and density of colonies declines steadily northwards and westwards. In Wales largely confined to areas on or near the N and S coasts. Also found on the Isle of Man, and occurs throughout Ireland but is very local and absent from large areas. Very rare in Scotland, with a 1908 record from Kircudbrightshire, but found on an old industrial site in E Ayrshire in 2003 and subsequently a few more scattered records.

There has been a modest decline in Britain, with sites destroyed by ploughing, spraying or, in the case of quarries and old pits, either filled with rubbish or overgrown by scrub. The losses have to a variable extent been compensated for by the colonisation of new sites. Bee Orchid is one of the more opportunistic and adaptable British orchids. There has been a much more significant decline in Ireland, especially in the period before 1930. The explanation for this is not clear.

DESCRIPTION

Underground The aerial stem grows from a pair of tubers. **Leaves** Basal leaves pale green, clearly veined, keeled and strap-shaped. Stem leaves narrower, more pointed and loosely clasping, with 1–2 bract-like non-sheathing leaves towards spike. Leaves appear September–November and are often scorched or otherwise damaged by the following summer. **Spike** Loose, with 2–7 flowers, sometimes as many as 12. **Bracts** Pale green, lanceolate, pointed and much longer than ovary. **Ovary** Green, boldly ribbed but not twisted, held upright but slightly curved; the tip bends further over to hold the flower facing outward. **Flower** Sepals oval, tapering slightly towards tip, concave and often hooded; various shades of pink, from pale rose to deep pink tinged lilac (occasionally white), with 3–5 variably obvious green veins; lateral sepals held horizontally or a little below horizontal and swept backwards, upper sepal upright but very frequently bent backwards to lie almost horizontally behind flower. Petals much shorter, strap-shaped with their margins rolled back; greenish through pinkish-brown to pink with fine white hairs. **Lip** Tongue-shaped with sides and front strongly moulded downwards and two relatively small conical sides lobes at base that are conspicuously hairy on outer side; tip with two lobes and a pointed nib in the shallow notch between them, but appears rounded as entire tip curled up underneath. At base of lip an elongated, semi-circular, hairless, dull orange area bordered by narrow maroon-brown and pale yellow bands. The speculum radiates from these, a broader band of dull purple in turn bounded by a pale yellow band. The markings form a U- or H-shape below the basal area, sometimes irregular and asymmetrical. Side-lobes also bounded by dull purple and pale yellow bands. Remainder of lip velvety maroon-brown. **Column** Greenish, held at *c.* 90° to lip (in profile it is said to resemble the head of a duck with the anther at the tip forming an elongated 'beak'). Pollinia yellow, their caudicles (stalks) lie in parallel grooves until they are released, with the viscidia at their bases enclosed in pale yellowish-green bursicles. Stigmatic cavity at base of column circular, yellowish with a horizontal band of orange-brown. **Subspecies** None.

sepals are occasionally whitish ➤

◀ Var. *chlorantha*

confused with old, faded flowers on normal plants.

Var. *trollii* Wasp Orchid Lip with a long narrow central lobe tapering to a point, side-lobes often longer and narrower than normal, and held away from central lobe. Lip marbled asymmetrically with yellow and rusty-brown, speculum either distorted or absent. Occurs regularly at a few sites in the West Country but is otherwise rare. Note that the tip of the lip in 'normal' plants sometimes fails to fold under and they then

VARIATION Flowers rather variable in structure, colour and pattern. As Bee Orchid is self-pollinated, any variations that occur spontaneously due to mutation can easily be passed to the plant's offspring to form populations of similar-looking variants. The same mutations seem to crop up from time to time in widely separated localities, and some of these are named varieties.

Var. *belgarum* Lip oval, lacking well-defined side-lobes but with hairy 'shoulders'. Has symmetrical markings with a horizontal yellow band across middle and smaller vertical yellow bands; no speculum. Scarce but widespread in S England.

Var. *chlorantha* Lacks anthocyanin pigments and has whitish sepals, yellow petals and a bright greenish-yellow lip with a 'ghost' pattern. Rare but occurs widely, especially in S and E. Var '*flavescens*' has also been described, a less extreme version of var. *chlorantha* with a pale brown lip that has a normal but faded pattern. Apparently very rare and easily

Var. *belgarum* ▶

Var. *trollii*

Var. *bicolor*

have an elongated, pointed lip but retain the usual markings. For a time, 'Wasp Orchid' was considered to be a distinct species.

Var. *fulvofusca* Lip unmarked, dark, chocolate-coloured. Very rare.

Var. *bicolor* Outer half of lip dark brown, grading to pale, unmarked greenish or pale brown at base. No speculum. Very rare but has been recorded from Dorset, Essex/Suffolk and Anglesey.

Var. *cambrensis* Described in 2014, has the lip paler and yellower with dark maroon confined to a spot near the tip. Known from Glamorgan.

Var. *badensis* Petals pink, enlarged and sepal-like (although not quite as big); lip normal. Very rare. Recorded from Suffolk, Dorset and especially Wilts.

Var. *friburgensis* Petals sepal-like, as var. *badensis*, but lip compressed (even dish-shaped in 'var. *saraepontana*'), elongated due to the failure of the sides and especially the tip to fold down and under normally, with paler ground colour, irregular markings, and the basal dull orange area broken. Very rare, Somerset.

Hybrids *O.* x *albertiana*, the hybrid with Late Spider Orchid, is recorded rarely from E Kent. ***O.* x *pietzschii***, the hybrid with Fly Orchid is very rare; recorded in West Sussex and at two sites N Somerset.

LATE SPIDER ORCHID *Ophrys fuciflora*

IDENTIFICATION

Rare and very local, confined to a few sites in the North Downs in E Kent between Folkestone and Wye. Height 5–30cm (–37.5cm). Unmistakably a 'bee' orchid, with pink sepals and petals and a broad, dark, velvety lip. The size and colour of the petals and the colour of the sepals are variable, as is the shape and pattern of markings on the lip, but usually has furry 'shoulders' and always a *projecting nib at the tip*. **Similar species** Bee Orchid has longer, narrower petals that are strap-shaped rather than horn-shaped and more often greenish than pink. It has a narrower lip that is distinctly 3-lobed and never square and 'shouldered' (on the other hand, the lip of Late Spider Orchid can be 3-lobed). Diagnostically, the pointed tip of the lip in Bee Orchid normally curls back and under out of sight and it therefore lacks a projecting nib. The column in Bee Orchid is a little longer with a slightly more prominent projecting 'beak', and the pollinia often hang loose over the stigma, a feature never seen in Late Spider Orchid.

Early Spider Orchid always has green sepals and long, narrow, strap-shaped petals that are much less downy and also usually greenish rather than pink. The speculum on its lip is normally H-, X- or π-shaped rather than incorporating broken rings and circles. As in Bee Orchid, its lip lacks a forward-pointing nib. The flowering periods of the two spider orchids do not normally overlap. **Flowering period** Late May–late July, exceptionally to August, but mostly early–mid June. At Wye NNR, colonies that face W–SW flower 3–4 weeks later than colonies facing south.

HABITAT

Well-drained grassland on infertile chalky soils, grazed to produce a reasonably short sward with some bare ground, and also ideally facing S. Ground disturbance of some sort may be important for the establishment of new populations. Current sites are on old spoil heaps and areas which were ploughed in the past or heavily disturbed by rabbits prior to the outbreak of myxomatosis. The individual plants are relatively long-lived, and colonies can persist when conditions are no longer suitable for seedlings to become established. Notably, the existing colonies tend to be very discrete and do not expand into adjacent superficially similar grassland.

POLLINATION & REPRODUCTION

Insect-pollinated. In Europe bees of the genus *Eucera* are the main pollinators, but the appropriate bees do not occur in England and insect-pollination has not been recorded, although it must occur – it is possible that pollen-beetles are a substitute, albeit far less efficient, for bees. Very few ripe seedpods have been found in Kent (and to compensate, hand-pollination has been undertaken at Wye). Fortunately, there is a very low level of adult plant mortality (less than 5% per annum), and only small numbers of seedlings need to survive to maturity in order to maintain the population, thus it seems that enough

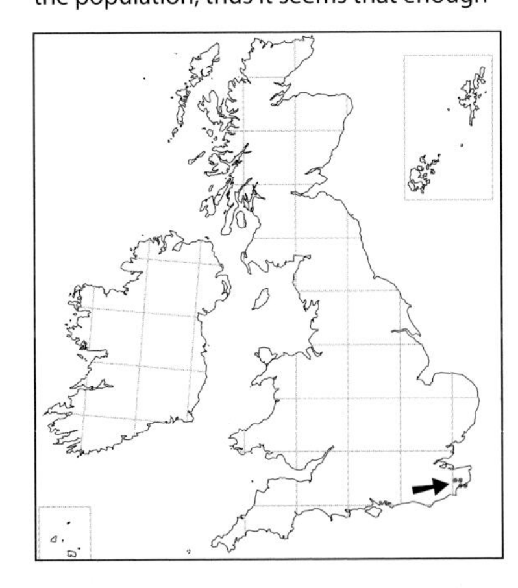

viable seed is produced. Self-pollination may occur occasionally, even in the bud.

Vegetative reproduction is either very rare or does not occur (very few new plants appear within 10cm, or even 30cm, of existing orchids).

DEVELOPMENT & GROWTH

Rather long-lived. In studies in Kent the 'half-life' of a population averaged 12.5 years and varied from 7.1–16.8 years (see p.251). Periods of underground 'dormancy' are frequent with *c.* 13% of plants dormant each season, and plants may spend 1–2 years underground before re-emerging.

There is no information on the length of the period between germination and the plant's first appearance above ground but it is probably 3–4 years. Many plants flower in their first year above ground.

STATUS & CONSERVATION

Nationally Rare and listed as Vulnerable: WCA Schedule 8. Only reliably recorded from Kent, with about 20 historic localities. Now much reduced, with plants appearing regularly at just five sites although there are probably still sporadic appearances at several more. The low point was in the mid 1980s, and there has been a substantial recovery since then, perhaps due in part to the weather but certainly to better management. There is now a total of *c.* 500 plants, almost half at Wye NNR (in six discrete colonies), where plants are caged for protection. The population there rose from *c.* 50 in 1987 to *c.* 220 in 1998. As with Bee Orchid, the number of flowering plants varies widely between seasons but the overall population is much more stable.

Habitat has been lost because of agricultural changes but also due to changes in grazing patterns and the reduction in rabbit numbers following myxomatosis, with grassland reverting to scrub. All the regular sites are now within SSSIs and are managed with the species in mind, but at its former localities grazing was abandoned many years ago. The Late Spider Orchid is probably best adapted to a 'dynamic' system. It needs ground disturbance and some bare ground to provide suitable conditions for seedlings to become established, as well as grazing and sometimes hand-mowing to maintain a suitable short sward for the adult plants.

DESCRIPTION

Underground The aerial stem grows from a pair of tubers. **Stem** Grey-green. **Leaves** Grey-green, prominently veined; 3–5 lanceolate basal leaves form a loose rosette, the lowest flat to the ground, the remainder held up to 30° above the horizontal. 1–3 narrower and more pointed leaves loosely clasp the stem. The rosette appears above ground in October and withers while the plant is still in flower, usually disappearing by early July. **Spike** 2–9 well-spaced flowers (rarely as many as 14). **BRACT** Grey-green, lanceolate, *c.* 1–1.5x length of ovary, which they clasp at the base. **OVARY** Grey-green, slim, cylindrical, boldly 6-ribbed, slightly twisted, also curved through *c.* 90° to hold the flowers facing outwards. **FLOWER** Sepals broadly oval, edges rolled back and under, tip blunt; very pale pink to a rich, dark pink with prominent green midrib on outer surface and 1–3 green veins on inner face. Lateral sepals are horizontally or slightly drooping at sides of flower, upper sepal vertical or projecting horizontally over the lip, but more often curving forwards in a graceful arch. Petals much shorter, triangular, velvety-hairy and pink (often distinctly deeper pink, sometimes almost flame-coloured, at base); may have swellings on each side at base. **LIP** Almost square in shape but usually broader towards tip. Edges strongly moulded downwards apart from tip, where a prominent forward or downward projecting nib set into a notch; nib usually yellowish and sometimes three-lobed. Lip rich, dark chestnut-brown or maroon-brown, usually paler around edges and velvety in texture. On either side of base of lip swellings or 'shoulders', variable in size: in some large, well-defined and rounded and thus more obviously lateral lobes. Lip particularly hairy on 'shoulders' and along sides below them. At base of the lip below column a more-or-less semicircular, smooth, orange-brown patch usually bordered by a narrow, creamy 'necklace'. Radiating from this necklace the speculum is a complex and extremely variable pattern of smooth maroon or lilac-brown markings bounded by creamy-yellow lines that often forms an irregular star or other geometric shape around a central dark circle. **COLUMN** Held at *c.* 90° to the lip, yellowish-green, becoming distinctly greener towards tip. Stigmatic cavity dark with a small, round, black 'pseudoeye' on either side. Projecting rostellum or 'beak' small, pollinia yellow. **SUBSPECIES** None. **VARIATION** Much variation in the shape and colour of the lip and colour of the petals and sepals. One named variety is worth mentioning: **Var. *flavescens*** lacks anthocyanin pigments and has whitish sepals and petals and a pale greenish lip with a very faint pattern. It is very rare. **HYBRIDS** ***O.* x *albertiana***, the hybrid with Bee Orchid, has been recorded rarely from E Kent. ***O.* x *obscura***, the hybrid with Early Spider Orchid, was found in Kent in 1828 (the parent species seldom occur together and the flowering periods do not normally overlap).

ADDITIONAL SPECIES

Summer Lady's-tresses
Spiranthes aestivalis

Extinct in Britain. It once grew in five bogs in the New Forest, Hants, as well as on Guernsey and Jersey in the Channel Islands, but the last confirmed record from the New Forest was as long ago as 1952 (with unconfirmed reports until 1959). It could be locally abundant ('half an acre of bog perfectly white with these flowers'), but the collection of specimens for herbaria, complete with rhizomes and roots, was a major factor in its disappearance; habitat change and habitat destruction were undoubtedly also involved. The flowers resemble those of Autumn Lady's-tresses but when in bloom mid July–mid August, there are serval long, narrow leaves at the base of the stem.

Sawfly Orchid ➤
Ophrys tenthredinifera

A single plant was found in April 2014 growing with Early Spider Orchids on the Purbeck coast of Dorset. A native of the Mediterranean (from Iberia east to Greece), it most probably grew from wind-blown seed. It did not re-appear in 2015.

Greater Tongue Orchid ▲
Serapias lingua

A single plant was found on Guernsey in 1992, but did not persist and was not seen subsequently. In 1998 three spikes were reported from near Kingsbridge in S Devon, and what was presumably a single plant lasted until 2003, when it apparently produced nine spikes; it has not been seen since. Unfortunately, the record was never substantiated by the publication of photographs and the plant was apparently seen by very few people, so the record must be regarded as unconfirmed. Separated from most other tongue orchids by the single solid dark purple boss at the base of the lip.

Small-flowered Tongue Orchid ➤
Serapias parviflora

Found near Rame Head in SE Cornwall in 1987, the plant (or plants) increasing in vigour over the years, but the last sighting may have been in 2008. It presumably originated from a single, wind-blown seed. The tongue orchids genus *Serapias* are essentially Mediterranean, but extend north to Brittany in France. Their taxonomy is very complex, with no agreement on the number of species, but Small-flowered has the smallest lip in the genus, the projecting tip just 6–11mm long.

GLOSSARY

achlorophyllose Lacking the green pigment chlorophyll and therefore unable to photosynthsise.
adventitious Buds and roots that appear in abnormal places on the stem.
ancient woodland Woodland that has maintained a more or less continuous cover of trees, probably for thousands of years.
annular Ring-shaped.
anther The pollen-bearing, male reproductive organ. In most orchids, the pollen is grouped into two pollinia.
anther cap In some orchids, such as the helleborines *Epipactis*, the anther lies on top of the column and is hinged or stalked. It may be contrastingly coloured.
anthocyanins Group of pigments that produce purple or reddish colours.
asymbiotic When a symbiotic fungus is absent.
autogamy Self-pollination with pollen from the same flower (see self-pollinate).
back-cross Cross between a hybrid and one of its parent species.
base-rich Soil with a high concentration of calcium or magnesium and a pH above 7.0.
bog Plant community on wet, acidic peat.
bosses Irregular swellings.
bract Structure at the base of a flower stalk, varying in size and shape, but often leaf-like.
bulbils Tiny, round growths, e.g. along the rim of the leaf of Bog Orchid, which can separate and are capable of developing into a new plant.
bursicle The pouch-like structure on the column of some orchids that contains and protects the viscidium (q.v.).
calcareous Rich in calcium carbonate, e.g. chalk, limestone or sea shells.
Caledonian woodland Ancient pine woodland, a relict of the 'Forest of Caledon' that supposedly once covered Scotland.
capsule The dry seed pod of an orchid.
carapace Hardened shell.
caudicle The stalk present in some orchids that attaches the pollinium to the viscidium (q.v.).
cilia Minute, thickened or fleshy hair-like structures.
ciliate With cilia projecting from the margin.
chlorophyll A green pigment, important in photosynthesis, found in discrete organelles (chloroplasts) in the cells of plants, usually in the leaves.
cleistogamy Self-pollination in bud, after which the bud may remain closed or may open.
clinandrium Depression on the top of the column, below the anther and behind the stigmatic zone, in which the pollinia lie.
clone Individual of identical genetic make-up to its 'parent' that results from asexual, vegetative reproduction.
column Specialist structure characteristic of orchid flowers in which the stamens and stigmas are fused together.
crenate With scalloped margins.
'Critically Endangered' Facing an extremely high risk of extinction in the wild in the immediate future.
cross-pollinate Pollination in which pollen from one flower fertilises another; usually taken to mean a flower on a different plant.
'Date Deficient' There is inadequate information on distribution and/or population status to make an assessment of the risk of extinction. More information is required and future research may show that a threat category is appropriate.
decurved Curved downwards.
deflexed Bent sharply downwards.
deltoid Shaped like the Greek letter 'delta', i.e. triangular.
diploid Having two matching sets of chromosomes. This is the normal state for plant cells (see tetraploid).
ectomycorrhiza Association of a fungus with the roots of a plant where the fungus forms a layer on the outside of the roots.

'Endangered' Facing a very high risk of extinction in the wild.
endomycorrhizal Association of a fungus with the roots of a plant in which the fungus penetrates the tissue of the root.
epichile Outer portion of the lip, often heart-shaped, in those orchid genera where the lip is divided into two (e.g. *Epipactis, Cephalanthera, Serapias*).
epidermis 'Skin' or surface layers.
epiphyte Plant growing on the surface of another plant without receiving nutrition from it.
esker Glacial debris, often sands and gravels.
fen Plant community on alkaline, neutral or very slightly acidic soil.
filiform Thread-like.
flexuous Wavy.
eutrophication Enriched with plant nutrients, often leading to a luxuriant and eventually stifling growth of vegetation.
geitonogamy Fertilised by pollen from another flower on the same plant.
glandular hair Short hair tipped with a small spherical gland containing oil or resin.
half-life A measure of the life expectancy of the orchid after its first appearance above ground, marking the point at which 50% of the population that emerged in any given year has died.
herbarium (plural: ***herbaria***) A collection of dried and pressed plant material.
hooded Formed into a concave shape resembling a monk's cowl.
hybrid Plant originating from the fertilisation of one species by another.
hybrid swarm Population in which the barriers between two species have largely or completely broken down. Hybridisation is commonplace and at random, producing a population that forms a continuous range of intermediates between the two parent species.
hybrid vigour When the first generation of hybrids between two species are exceptionally large and robust.
hyperchromic Intensely coloured, with an excessive amount of pigmentation.
hyperresupinate When the ovary and/or flower stalk twist through 360° to position the lip at the top of the flower, e.g. Bog Orchid.
hypha (plural: ***hyphae***) Fine, thread-like structures that make up the body of a fungus.
hypochile Inner portion of the lip, often cup-shaped, in those orchid genera where the lip is divided into two (e.g. *Epipactis, Cephalanthera, Serapias*).
intergeneric hybrid A hybrid whose parents are in two different genera.
lanceolate Narrowly oval, tapering to a more or less pointed tip.
lax Loose, not dense.
Leblanc Chemical process used to produce washing soda.
lip Highly modified third petal of an orchid; also known as the labellum.
lough Irish term for loch.
machair Sandy, lime-rich soil with a species-rich sward of short grasses and herbs. The machair is confined to the coasts of W Ireland and W Scotland.
meadow Grassy field from which livestock are excluded for at least part of the year so that it can be cut for hay.
monocarpic Flowering once and then dying.
mutualism An intimate relationship between two or more organisms from which all derive benefits.
mycorhizome Early stages in the development of the underground rhizome in which the seedling is nourished entirely by fungi.
mycorrhiza Association of a fungus with the roots of a plant in which the fungus may form a layer on the outside of the roots (ectomycorrhizal, q.v.) or penetrate the tissue of the root (endomycorrhizal, q.v.).
mycelium The mass of branching filaments that make up the body of a fungus.
mycotrophic Acquiring nutrition from fungi.

'Nationally Scarce' Occurring in 16–100 hectads (10km x 10km squares) in Great Britain.
'Nationally Rare' Occurring in 15 or fewer hectads (10km x 10km squares) in Great Britain.
native Growing in an area where it was not introduced, either accidentally or deliberately, by humans.
non-sheathing leaf A leaf with its base clasping the stem but not completely encircling it.
NNR National Nature Reserve.
ovary Female reproductive organ that contains the ovules.
ovule Organ inside the ovary that contains the embryo sac, which in turn contains the egg.
pH Measure of acidity.
parasitic Organism that lives on or at the expense of other organisms.
pasture Grassland that is grazed for some or all of the year but not cut.
petals Inner row of 'perianth segments', one of which is modified to form the lip.
papilla (plural: ***papillae***) Small, nipple-like projection.
pendant Hanging downwards.
peloton A coil-like structure formed by fungi inside the cells of an orchid.
pheromone Chemical secreted by an animal, especially an insect, that influences the behaviour or development of others of the same species.
photosynthesis Production of food by green plants. In the presence of chlorophyll and light energy from the sun, carbon dioxide and water are converted into carbohydrates and oxygen.
phototrophic Acquiring nutrition through the process of photosynthesis.
pollen Single-celled spores containing the male gametes.
pollinium (plural: ***pollinia***) Regularly-shaped mass of individual pollen grains which is transported as a single unit during pollination; the pollinia are often divided into two.
propagules Various vegetative portions of a plant such as a bud or other offshoots that aid in dispersal and from which a new individual may develop.
protocorm Initial stage of development for every orchid formed by a cluster of cells.
pseudobulb Swollen or thickened portion of stem, covered in the leaf bases. It fulfils the same storage function as a bulb or tuber; found in Fen and Bog Orchids and common in tropical species.
pseudocopulation Attempts by an insect to copulate with an insect-mimicking flower.
pseudopollen Structures in a flower that imitate pollen in order to attract insects, e.g. in the *Cephalanthera* helleborines.
reflexed Bent back or down.
resupinate When the ovary and/or pedicel twist through 180° to position the lip at the bottom of the flower.
reticulation Marked with a network of veins.
rhizoid Hair-like structure on the surface of a protocorm or mycorhizome that facilitates the entrance of fungi (i.e. a root hair).
rhizome Underground stem that lasts for more than one year from which roots and growth buds emerge; also a horizontal stem, either growing along the surface or underground.
rostellum A projection from the column that often functions to separate the pollinia from the stigma and thus prevents self-pollination (the rostellum is actually a modified sterile third stigma). The rostellum may exude a viscidium (q.v.).
runner Horizontal stem that grows along or just below the surface of the soil.
saprophytic Plants, fungi, etc. that feed on dead organic matter.
scale leaf A leaf that is reduced to a small scale.
secondary woodland Woodland in which the continuity of tree cover has been broken for a substantial period of time.
self-pollinate Pollination of a flower by pollen taken from the same plant; usually used in the context of autogamy, where

pollination is by pollen from the same flower, but sometimes also geitonogamy, where the pollen comes from another flower on the same plant.
sepal Outer row of 'perianth segments' that form the protective covering of the bud. In orchids they are either green or brightly coloured and form a conspicuous part of the flower.
symbiosis An intimate relationship between two or more organisms; formerly used for a relationship where all participants derive benefits (mutualism q.v.) but now used in a broader sense to include parasitism.
sheathing leaf Main leaf on an orchid with a base that completely encircles the stem.
sinus Indentation between two lobes on the lip, used especially when describing the flowers of marsh orchids and spotted orchids in the genus *Dactylorhiza*.
speculum Pattern on the lip, often with a metallic lustre or shine, in the bee and spider orchids, genus *Ophrys*.
spur Sac-like extension of the base of the lip which contains nectar in some species of orchid.
stamens Male reproductive organs of a flowering plant.
staminode Sterile stamen that forms a prominent shield-shaped structure within the flower of Lady's-slipper.
stigma Receptive surface of the female reproductive organs to which the pollen grains adhere.
stolon A horizontal stem, either growing along the surface or underground, not necessarily forming a new plant at the tip.
sympodial Pattern of growth in which the tip of the stem or rhizome either terminates in a flower spike or dies each year. Growth continues from buds formed at the base of the old stem.
synsepal Structure formed when the two lateral sepals are joined for almost their entire length, found in Lady's-slipper.
tetrads Group of four pollen grains originating from a single 'mother' cell.
tetraploid Having four sets of chromosomes (see diploid).
tubers Swollen underground roots or stems, functioning as storage organs.
turbary An area where peat or turf is cut for fuel.
turlough Irish term, used for a seasonal lake on limestone in which the water level may fall dramatically in summer.
vice-county Divisions of Britain and Ireland based on the 19th century county boundaries that are used for biological recording; larger counties, such as Norfolk and Yorkshire, are divided up into smaller units.
viscidium (plural ***viscidia***) Detachable sticky exudation from the rostellum that attaches the pollinia (sometimes via a short stalk, the caudicle) to a visiting insect.
'Vulnerable' Considered to be facing a high risk of extinction in the wild.
WCA Schedule 8 Wildlife & Countryside Act Schedule 8, which lists all the wild plants that are specially protected, making it illegal to damage or destroy them in any way (including removing flowers or leaves, or even seed; see p.17).

INDEX OF SPECIES

Autumn Lady's-tresses ➤